Petroleum, natural gas and coal

Nature, formation mechanisms, future prospects in the energy transition

Bernard Durand

edp sciences

EDP Sciences
17, avenue du Hoggar
Parc d'activités de Courtabœuf, BP 112
91944 Les Ulis Cedex A, France

Printed in France

ISBN (print): 978-2-7598-2231-7 – ISBN (ebook): 978-2-7598-2232-4

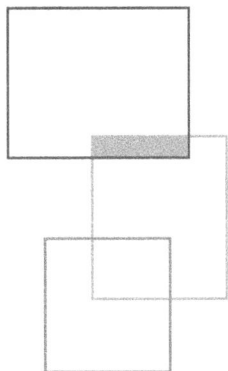

Contents

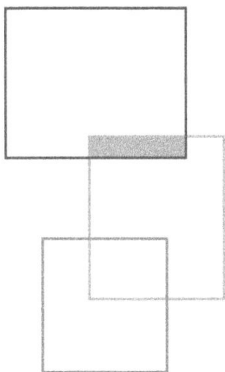

The author

Bernard Durand is a fossil fuels geochemist, former head of the Géologie-Géochimie Department of the Institut français du pétrole et des énergies nouvelles (IFPEN) and former director of the École nationale supérieure de géologie (ENSG). Alfred-Wegener Award of the European Association of Geoscientists and Engineers (EAGE).

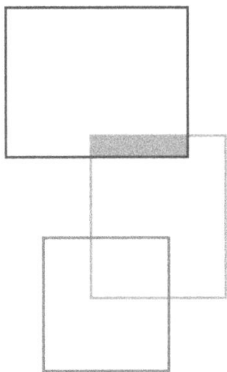

Preface

We love fossil fuels. We really do. Otherwise, would we spend between 300 and 700 billion dollars each year to find new oil and gas fields, and enhance the capacity to extract what they contain? Would we protest loudly each time the government considers to raise taxes on gasoline or domestic fuel oil? Would countries have sent armies to conquer or protect coal mines, then oilfields, a number of times in the course of recent history? Would Hitler have invaded Russia if it were not for the Caucasian oil fields, and would Japan have bombed Pearl Harbor if it were not to get rid of the US marine and secure a naval route to the Indonesian oil fields?

Actually, we love fossil fuels so much that it is acceptable to say that we are addicted to them. We can plea to have good reasons to do so, though: their use gave us all the pleasures we could dream of. Through fossil fuels, we have acquired the freedom to move fast, on the Earth, on the sea or in the air. We got heat in the winter and cool air in the summer. We got electricity (70% of the world production comes from coal, gas and oil), which in turn gave us all the domestic comfort, and the plants that manufacture all that we can find in a store.

Fossil fuels also tremendously increased our food production, through the N-fertilizers made from gas, and the agricultural machinery running on oil. Do you know that when you eat a kilogram of beef, you kind of eat a kilogram of fossil fuels, that were used to grow the cereals eaten by the cattle?

They gave us organic chemistry then plastics, steel and cement, and thus all the buildings that have appeared in the 20th century, but also clothes (all synthetic fibers are oil and gas derivatives), shoes, windows and toys, coffee machines and carpets, beds and fridges, TV screens and chairs, skis and shampoo, forks and planes, coffee cups and dishwashers, and of course smartphones and modern communications.

They have also made possible to install everywhere water distribution systems and sewage, that, together with the increase of food security, have played a major role - far greater than curative medicine - in tripling our life expectancy between 1800 and 2000.

Actually, fossil fuels gave us everything we see in the modern world. Suppress them, and we lose computers thus money, transportation thus food – and all the other goods – in the cities, and the productivity of work, that is merely empowering each of us with machinery that multiplies our mechanical power by hundreds or thousands.

The list is not finished: because they allowed a tremendous increase in the productivity of work, fossil fuels also gave us long studies, vacations and retirement, they enabled to put 80% of the population in cities (because we do not need so many people in the fields, where they have been replaced by tractors, pesticides and fertilizers), and they have allowed globalization – because of boats, trucks and planes – and modern financial markets (that require computers and electricity).

And all this, nature has given it to us for free. What, fossil fuels would be free? Indeed, they are. Just as no one has ever paid a single cent for wind to exist, no one has ever paid a single cent for oil – or gas, or coal – to exist. This precious resource has been formed by Mother Nature without the help of any human, with remnants of ancient life, for hundreds of millions of years. Actually, when the process of oil formation began, there were not that many humans around to give a helping hand!

The price of oil, today, is only money paid by some humans to other humans: those that have worked to extract it from the environment, those that have been lucky enough to sit on the oilfield, and all those that transport it or trade it. But not a single cent ever went to Nature, that manufactured it.

And it is precisely because it is so much easier to extract energy from the environment – and store it, and transport it, and use it whenever you want – when it is coal, oil or gas than when it is wood, wind or the run of water that humanity has moved from a 100% renewable civilization to the present state. It made energy much more available, both in the physical and in the economic sense.

We have not become addicted to fossil fuels just because we are total idiots. We have done so because these energies can be used whenever we want and not just when the sun shines or the wind blows, or just where there are forest and mountains (for water mills then hydroelectricity). They have enabled to use energy elsewhere than where it is harnessed. Wood can be stored but not easily moved around and try to store running water or wind!

So, if these fuels have made easy for all, what is the point of bothering about the future? Every coin has two faces, and these energies have alas two disadvantages: they are subject to depletion, and they cause climate change. Depletion is easy to understand for a non-renewable resource. It's the application of "you cannot eat your cake and have it". Once you have extracted and burnt a resource that takes several ten million to several hundred million years to renew, you have less.

Mathematics allows to demonstrate that when you draw from a stock that has been given once and for all, all you can get is a yearly production that starts at zero (which has been the case in the remote past), will eventually fall to zero, and go through an absolute maximum in between, named the peak. This is not limited to oil or gas, though: it is valid for any metal ore, or phosphates, or potash. The only question, so to say, is when the peak might occur (and should we trigger it for environmental reasons or wait for it to happen for other reasons), at what level, and with what consequences. The oil production of the North Sea peaked in 2000, and the world production of conventional oil (everything except tar sands and shale oil) peaked in 2006, so this is no virtual process!

And once you have burnt a fossil fuel, mostly made of carbon because it's ancient life, you have created carbon dioxide, because combustion is a particular form of oxidation. And, because any oxide is a stable molecule, carbon dioxide cannot be removed fast from the atmosphere, and accumulates in the air, trapping more infrared close to the ground and causing global warming. This process has been identified almost two centuries ago by Joseph Fourier, it's solid science!

What is also solid science being that, when it went from the last ice age to today, the planet warmed by only 5 degrees (Celsius). So, a global warming of a couple degrees in a century is likely to trigger war everywhere, because our sedentary species will be unable to adapt to such a rapid and massive change in the conditions that have prevailed for millennia and framed the civilizations everywhere.

So, what is at the heart of our modern world is at the root of a Faustian pact: with oil and its friends, we can have the land of plenty today, but once we are totally dependent on it, at some point we will have less, and on top of that we'll pay a high price in terms of global environmental issues.

Bernard Durand has not the ambition to suggest an easy way out in this situation. But he has one, for sure!

From the beginning of this foreword, I have used the words oil, gas, coal, and "fossil fuels" without defining them. Easy, you will think? Well… Did you know that part of what is counted with oil comes from gas fields, and vice versa? That nobody knows exactly how much oil we can extract from the ground, because there is no centralized information on the topic?

That oil production is counted in barrels, which is a unit of volume, and not in kWh or joules, so that nobody knows exactly what is the energy content of what is extracted? That the proven reserves of the Middle East countries were worth 360 billion barrels in 1980, and 810 at end 2016, with 280 billion barrels extracted – or "produced" – in between, and no major discoveries in that time interval?

So, before trying to answer the questions "how long can we extract enough fossil fuels from the earth's crust to sustain our present way of life, and "how can we do with less", because we will have to do so one day, Bernard just wants to enable the reader to understand what we are talking about: What are fossil fuels exactly, and what are they made of? How were they formed in the Earth crust? How do oil and gas companies know that there are hydrocarbons under the ground somewhere? Is there a common way to define reserves, production, and consumption, or everyone has his own way to provide a figure? When the issue is crucial, how do we know what we know, or why can't we know what we would like to know? My dear reader friend, do not imagine a single second that you can avoid to go through the pain of learning, because the mighty people "know all this, and take it into account when framing their decisions". I have discussed with 5 or 6 governments in France since 2003, and dozens of members of parliament in my country. Trust me: when it comes to technical topics – and energy is a lot of technique – they know nothing more than the ordinary citizen.

So, Bernard has a tremendous ambition: he is still dreaming that, if he has enough readers, it might enhance the knowledge of enough voters and enough decision makers to change a little something to the future. My dear reader friend, I hope you will enjoy understanding better what are those fossil fuels that have made possible all your present living conditions, and I wish the best success to Bernard and to this book.

Jean-Marc Jancovici

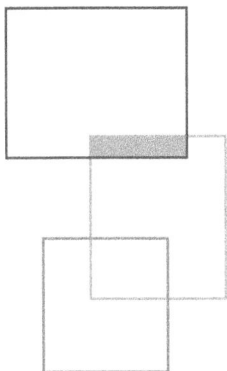

Foreword

"As long as the weather is fine, the man does not predict the storm"

Niccolo Machiavelli

For more than a century now, carbonaceous fossil fuels (petroleum, gas and coal) have provided the bulk of the flow of energy that gives life to industrial societies.

In these societies, therefore, a strong link existed between the increase in their consumption of fossil fuels and their material and human development. But, perhaps by virtue of the adage "don't look a gift horse in the mouth", their public opinions and most of their decision-makers did not have a clear conscience of this link, and still do not seem to really perceive it. Even now, carbonaceous fossil fuels are the forgotten ones of the debate on the energy, which invades the public sphere and the media. Indeed, this debate mainly deals with electricity, renewable against nuclear energy.

This period of recklessness will not last very long: Questions of the professional circles on the future availability of carbonaceous fossil fuels are beginning to reach the opinion. The latter also seems to better realize their importance, that of oil in particular, in everyday life. It is also more and more concerned about the climate change caused by the greenhouse gas emissions resulting from their use.

Undoubtedly, the industrial countries will be, are already, facing a major and even existential problem: that of the passage of human societies based on the massive use of fossil fuels to societies that have learned to do without. This will be, by necessity, the real engine of the energy transition that we are talking about so much now! For it is the future availability of fossil fuels that will give the tempo of this transition, not the development of other sources of energy, so difficult is the way they still have to do to replace fossil fuels!

Only a few technical developments on the methods used by industry to find, extract and process fossil fuels can be found in this book. There will also be no detailed statistics of their production and consumption, nor a treatise on economics. Its objective is to provide non-specialists, first and foremost those preoccupied, or concerned in their activity, by energy transition and climate protection, with basic knowledge enabling them to better understand the nature of fossil fuels, their importance in the economies of the industrialized countries, and why their future availability will play an essential role in the future economic and social transformations in these countries.

It comprises two parts, which can be read widely independently of one another:

- The first part has a pedagogical purpose: in a few pages and avoiding too technical developments, it is a question of presenting an inventory of the variety of fossil fuels and of understanding the physicochemical principles that govern their formation and that of their deposits in the earth's crust.

- The second part is more speculative: on as quantitative bases as possible, we discuss major current issues: What are the remaining reserves of fossil fuels? When precisely during this century can we foresee the decline of their productions? What consequences will this decline have for industrial societies? What is and will be their role in climate change? What risks does their use entail for public health?

On the road is specified what are categories of fossil fuels such as oil and gas known as shale oil and shale gas, oil shales and tar sands, lignite and coal, conventional and unconventional oil and gas... which are too often the subject of passionate debates without that their exact nature is well understood.

I would like to thank most warmly for the help they have given me in the writing of this book, by providing me with essential documents or attentive criticism: Pierre Alba, Denis Babusiaux, Pierre-René Bauquis, Art Berman, Dorothy Bjorøy, François-Marie Bréon, Patrick Brocorens, Xavier Chavanne, Hubert Flocard, Marcel Descamps, David Fridley, Jean-Marc Jancovici, Jean Laherrère, Claude Laffont, Michel Lepetit, Jean-Marie Martin-Amouroux, Luis Martinez, Euan Mearns, Nicolas Meilhan, Matt Mushalik, Christian Ngô, Hervé Nifenecker, Christian Ravenne, Olivier Rech, Alexandre Rojey, David Rutledge, Jacques Treiner and Roland Vially. Without them, this work would have been impossible.

I am specially indebted to Jean Laherrère, whose outstanding work on the subject was a source of inspiration, and who also provided me with invaluable graphs and data.

But of course, the interpretations and conclusions of this book are of my own.

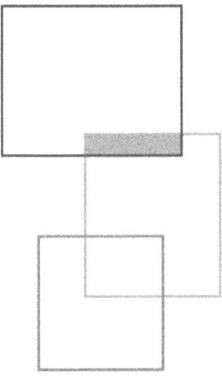

Introduction
The importance
of fossil fuels
for industrial societies

Living organisms cannot exist and develop without the energy provided by their food. Similarly, human societies cannot exist and develop without the energy they take from natural sources, which is called primary energy.

The quantity of primary energy, which these societies require, is closely connected with the nature and quantity of the material objects, which they produce. In fact, these material objects can be produced only by transforming raw materials, and no transformation of matter is possible without energy.

More than 80% of the world's primary energy currently comes from fossil fuels. That is to say the dependence on these of human societies, and especially the industrial societies, which are the most consuming of them!

Global consumption of fossil fuels has grown dramatically since the mid-19th century: per capita of the planet, it was then multiplied by about 30 (Figure 1). But in the meantime, the world's population has been multiplied by just over 6. The world's consumption of fossil fuels has thus been multiplied by about 200!

This growth took place in three successive waves: first coal, then oil, and then natural gas.

It is observed on Figure 1 that as a whole, periods of exponential growth of primary energy per capita alternate with periods of moderate growth. Moreover, since 2013

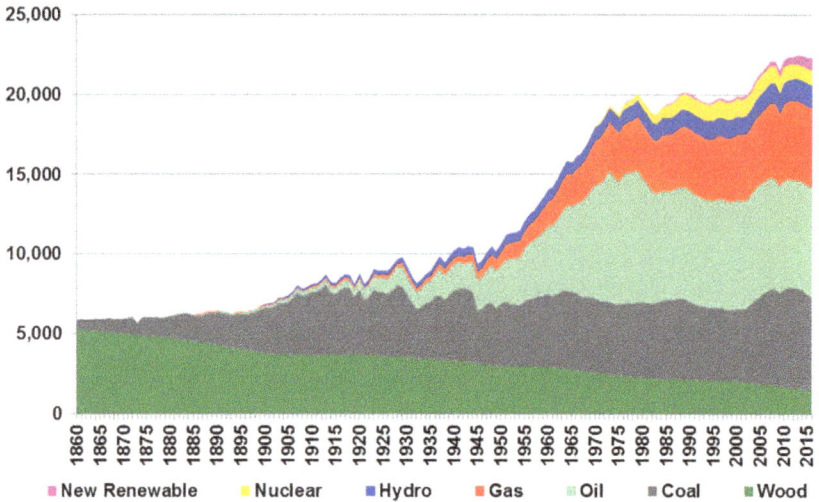

Growth of world primary energy consumption from 1860 to 2016, in
kWh per capita, and breakdown by primary energy source.
Courtesy JM Jancovici

■ New Renewable   ■ Nuclear   ■ Hydro   ■ Gas   ■ Oil   ■ Coal   ■ Wood

Figure 1	*Evolution of world consumption of different sources of primary energy from 1860 to 2016, in kWh per capita. Fossil fuels, oil, coal and natural gas, by order of decreasing importance, account globally for just over 80% of this consumption in 2016. "New Renewable" refers to so-called renewable energies, other than firewood and hydroelectricity. The kilowatt hour (kWh) is an energy unit worth 3.6 million joules (MJ). It is customary to use it to measure quantities of electrical energy, but nothing prevents it from being used for other energies. The joule (J) is the energy unit of the International System of Units (SI), but it is a very small unit: to raise the temperature of one liter of water from 0 °C to 100 °C, approximately 418,500 joules (418,5 kJ) of heat are required. More figuratively, it takes a mechanical energy of 1 joule to raise by one meter a small apple (with a mass of 101.94 grams).*

the total primary energy per capita is declining, this being mostly due to a decline of coal consumption in China.

The most consuming countries are of course the industrialized countries, of which China is now a part. The proportions of the different fossil fuels used vary widely from country to country (Figure 2), between 50 and 100% of their primary energy consumption, excluding firewood.

Of all these major consuming countries, France, with roughly 50% of its primary energy consumption, excluding firewood, uses the least fossil fuels in its energy mix[1]. This is due to the importance of its nuclear power production. Saudi Arabia uses them most, 100%.

---

[1] The energy mix of a country is the set of energy sources it uses, detailed by percentage of each source in total consumption.

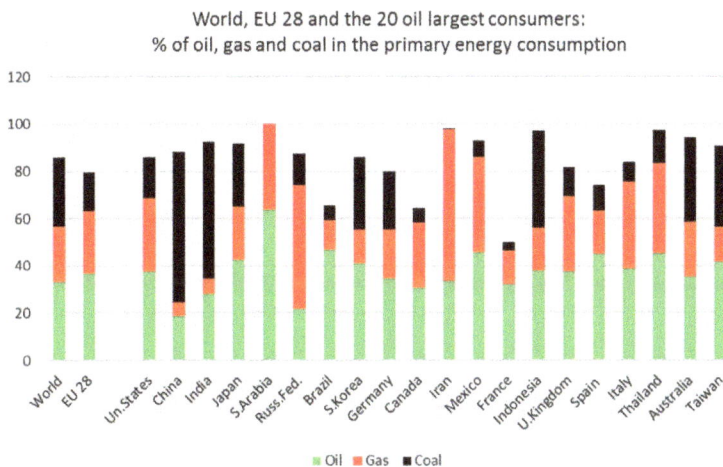

World, EU 28 and the 20 oil largest consumers:
% of oil, gas and coal in the primary energy consumption

Figure 2	*Proportions of fossil fuels in percentage of the primary energy consumption of the world, of EU 28 and of the 20 most oil-consuming countries, ranked from left to right by decreasing importance of their oil consumption. These data come from the British Petroleum Statistical Review 2016, which does not take into account traditional biomass (firewood...). Taking this into account would reduce the world average share from 86% to about 81%, the most frequently cited value, but would not change much for the highly industrialized countries, which are relatively non-wood consumers.*

Still excluding firewood, the world average was 86% in 2015. For the United States, the proportion was also 86%, for China 88.2% and for India 92.5%. For the EU28, despite its efforts to develop recourse to other sources, it was still 80%, and for Germany, although champion of renewable energies, also 80%.

Now, there is a strong correlation between growth in primary energy consumption and economic growth (Figure 3). Fossil fuels, which provide the bulk of primary energy in all industrialized countries, have been and continue to be the main engines of their economic development.

This growth in the per capita primary energy supply of the planet has brought about tremendous changes in the organization and lifestyles of human societies, first of all in the countries which gave birth to the Industrial Revolution, initially fuelled by coal, because of the considerable increase in labor productivity thus permitted. This has led to major social advances (material abundance, health, education, hours of work, reduction of social inequalities, etc.), as well as serious conflicts and human tragedies, and great inequalities in the development of Countries according to their unequal access to fossil energy.

This growth has also been accompanied by a sharp increase in the world population, driven by medical advances and increased agricultural yields, firstly in industrialized countries, then in many of those which are not or not yet. Most of the men currently living are somehow the children of fossil carbon energies.

World, EU 28 and the 20 most oil-consuming countries in 2015: Variation, in %, of their primary energy consumption from 2002 to 2015. Source : BP

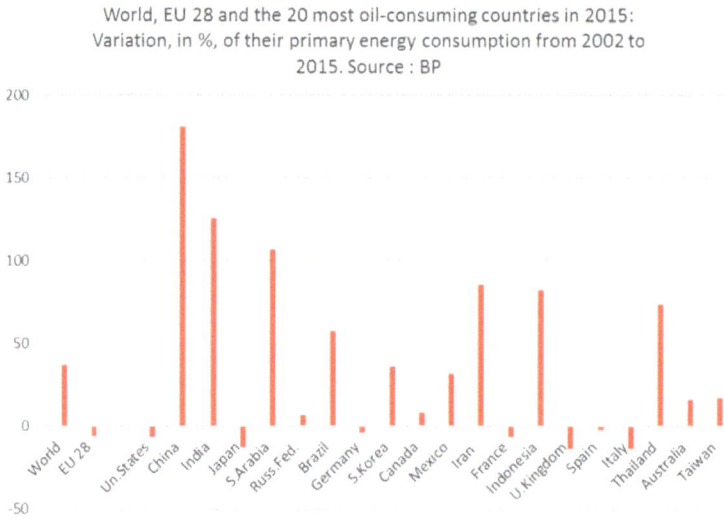

Figure 3   *Change in the primary energy consumption of the 20 most oil consuming countries during the period 2002–2015, ranked from left to right by decreasing importance of their oil consumption. The countries with strong economic growth are those whose primary energy consumption has increased significantly, with the most spectacular growth being China. In negative values are those who have experienced a decrease in this consumption: they are also those who have experienced a decline or a stagnation of their economy during this period. Source: British Petroleum Statistical Review 2016.*

But the global consumption of fossil carbon energy has now become such that its growth seems very difficult to maintain during this century, first for oil and natural gas: when looking at Figure 1, it should be noted that after a nearly exponential increase, the rate of consumption has a tendency to decrease with time. Greenhouse gas emissions and the nuisance caused by their use are also of concern to those who care about climate and public health.

An energy transition, the transformation of human societies that have based their development essentially on fossil fuels into societies capable of doing without them, is therefore inevitable. However, other known primary energy sources, nuclear energy and "renewable energies" are currently only marginally replacing them on a global scale. Without doubt, therefore, the future availability of fossil fuels will give the tempo of this transition. Yet, very surprisingly, they have barely been mentioned in the debates on the Energy Transition in France.

It would seem, therefore, that in France and probably in other industrialized countries too, economists, politicians, media and even teachers and non-specialist scientists, influencing public opinion, have not yet taken the real measure of the situation.

Obviously, in order to understand the problems involved, a good knowledge of what fossil fuels are really about and their future production possibilities is necessary. The purpose of this book is to provide the reader with the basics of such knowledge.

Its first part consists of a description of the variety of fossil fuels, of the principles of their formation, and of the physical mechanisms of the formation of their deposits, i.e. their economically exploitable accumulations. This description, although not very technical, may seem austere to a non-scientific reader. Its reading is nevertheless useful for those who want to acquire the basic vocabulary and an overall vision. In its second part, quantitative benchmarks are given, particularly on their remaining reserves and production forecasts over the course of this century, and then the relationship between fossil fuel consumption and climate and the risks of their use for human health.

Explanations of categories of fossil fuels, such as conventional and unconventional petroleum and gas, shale oil and shale gas, oil (bituminous) shales and tar sands, or lignite, the exact nature of which often does not seem to be well understood, are also given.

The accumulations of fossil fuels, despite their wide variety, all originate from the incorporation into some sediments of significant amounts of biologic debris and biologic substances, the whole of which constitutes the so-called kerogen. These debris and substances are the remains of organisms that have lived in the sedimentation environment or on neighboring land. Kerogen is then transformed during geological times by physicochemical mechanisms linked to the progressive burial of the sediments that contain it, giving birth to fossil fuels and their deposits.

This organic origin was fully recognized for coals as early as the middle of the 19th century, when the debris of plants from which they were mostly formed could be commonly identified under the microscope. On the other hand, it was not until the 1960s for natural oils and gases, because they did not contain a visible marker of their origin, even under a microscope. The methods of investigation had still been insufficient to carry a unanimous conviction. Decisive observations were then made. They have been fully confirmed over the last 50 years by the practice of oil and gas exploration.

A difficulty in understanding the subject is inherent in geological studies: time scales are far superior to those governing human life. The formation of fossil fuel deposits is an incredibly slow phenomenon on a human scale. It takes place in immense natural physicochemical reactors, sedimentary basins.

This very low formation rate makes it impossible to carry out laboratory experiments in order to faithfully reproduce the mechanisms. It was thanks to an intense observation, measurement and modeling work that we could really understand them.

Computer simulation models of the formation of oil and gas deposits in sedimentary basins during geological time have now become commonly used by oil companies to guide exploration. Some go a long way in the details and are of a complexity that has nothing to envy to that of the climate models that currently hold the spotlight. The most powerful computers available on the market have been used since the 1980s to exploit these models.

First part

Nature and variety of fossil fuels, physicochemical principles of their formation and that of their deposits in the earth's crust

1

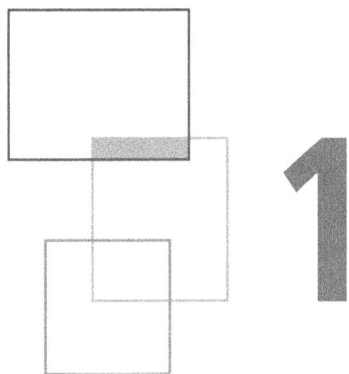

What are the fossil fuels, and what are they made of?

Fossil fuels are fluids or solids found in the earth's crust, and more precisely in sedimentary basins, that is, depressions formed during geological time and invaded by water, where erosion products (sediments) have accumulated from neighboring continents. To be exploitable, they must be present in large enough accumulations (deposits) and be easy to access in order to have an economic interest. Fluids are natural oils and gases, solids[2] are mainly coals and bituminous (oil) shales. We use these terms deliberately in the plural, given the wide variety of fossil fuel composition in their deposits.

1.1 The oils

They are made up for the most part of hydrocarbons, that is to say molecules containing only carbon and hydrogen.

---

[2] Solids include also methane hydrates (methane ice) found in permafrost in arctic areas, or in ocean and lake sediments by water depths of a few hundred meters, of which there are very large quantities: they are combinations of molecules of water and molecules of methane, in the solid state and stable under these conditions, but which are not any more when brought to the surface. At present (see Chapter 3-1-5), their reserves, i.e. the quantities that can be extracted under economic conditions, are insignificant. Methane, of formula CH_4, is the simplest hydrocarbon, which are molecules formed solely of carbon and hydrogen.

Hydrocarbons of oils are liquids under so-called normal conditions of temperature and pressure (in the petroleum industry, 15 °C and atmospheric pressure). They are classified into structural families (see Appendix 1). Their number of carbons, that is to say the number of carbon atoms contained in their molecules, varies between about 5 and 30, with a mode generally in the region of 10, but can range up to 50 and more! The result is a number of possible different combinations of carbon and hydrogen, that is to say, distinct molecules of hydrocarbons, which is extremely high, of the order of several billion!

With hydrocarbons, petroleum also contains sometimes a high proportion of other types of molecules, and this is little known to non-specialists, which combine carbon and hydrogen with other atoms (heteroatoms), sulfur, oxygen and nitrogen mainly. These heteroatomic molecules have a great influence on petroleum properties. They are very diverse: some have low molecular weights, including thiols also known as mercaptans, which contain sulfur, and phenols and furans, which contain oxygen. They are often volatile and contribute to give a nauseating odor to many crude oils. In medium molecular mass molecules, there are benzothiophenes and dibenzothiophenes, in which one ring of four carbon atoms and one sulfur atom is attached to one or two benzene rings (also called aromatic rings, see Appendix 1). But the essential consists of molecules of great molecular mass. These "heavy" molecules are for the most part impossible to study individually, and even in fractions (an exception is that of porphyrins, see Appendix 1). Only their overall properties are accessible, elemental analysis, average molecular mass, physical properties, etc.

Called resins and asphaltenes, they can constitute a very important fraction of the oils (about 15% for a current oil). They are also called NSO, for nitrogen (N), sulfur (S) and oxygen (O). The distinction in the laboratory between resins and asphaltenes is empirical: asphaltenes precipitate when a light saturated hydrocarbon, usually pentane C_5H_{12} or heptane C_7H_{16}, is mixed with a crude oil sample, while the resins remain in solution. However, this corresponds to overall structural differences: the structure of the resins allows them to be soluble in hydrocarbons, and to keep the asphaltenes in pseudosolution (see Appendix 1). Precipitation can occur naturally if light hydrocarbons invade a deposit of an asphaltene-rich oil in its geological history; this is called deasphalting.

There is therefore not one petroleum, and the variety is very great. They are classified summarily according to their density in relation to pure water, in light, medium, heavy and even extra heavy oils. This density varies according to their composition, between about 0.75 and 1 and can sometimes be slightly less than 0.7, or reach a little more than 1. Generally speaking, the heavier the oil (denser) the greater the content of resins and asphaltenes.

In the petroleum literature, dominated by North American habits, this density is expressed in API (American Petroleum Institute) degree, American variant of the Baumé degree, calculated by the formula Degree API = (141.5/(density at 60 °F)) – 131.5). Table 1 shows the correspondence between API degrees and density with respect to water. It can be seen that the API degree is the inverse of the density.

| Table 1 | Correspondence between density with respect to water (D) and degree API. |

D	1.014	1.007	1.000	0.966	0.934	0.904	0.876	0.850	0.825	0.802	0.780	0.759	0.757
API	8	9	10	15	20	25	30	35	40	45	50	55	58

The most common oils (medium) have an API degree of between 25 and 35.

Shale oil, whose exploitation has seen spectacular development in the United States after 2010, thanks to technological progress and rising oil prices, is a light oil of degrees API varying from about 35 to about 45. It has no other characteristic than to be in very low permeability reservoirs located in or in contact with the rock where it was formed, its source-rock (see Chapters 2.3, 4 and second part). Hence the name it is given more and more is source-rock oil, and most frequently light tight oil (LTO).

Petroleum with an API degree of less than or equal to 10, i.e. denser than pure water, is referred to as extra-heavy (XH). They include extra-heavy oils, and bitumens (bitumens, tars), also known as asphalts (asphalts), which are highly viscous liquids. Bitumens can be even in the solid state.

The distinction between extra-heavy oils (extra-heavy oils, XH) and bitumens (bitumens, tars) is based on their viscosity when they are in place in the field. This is very influenced by the temperature of the deposit. Bitumens are classified as having a viscosity greater than 10,000 centipoises (cP). Of course, the viscosity of bitumens and extra-heavy oils in their fields is of great importance for the choice of their methods of exploitation.

The most important fields of bitumen are those of Athabasca in Canada, in the form of so-called tar sands. These are huge fields found on the surface or at shallow depths (less than 400 meters) over very large areas, about 60,000 km^2. Their average temperature is about 5 °C and their viscosity in the order of 10^6 cP.

The most important fields of extra-heavy oils, of as great importance as the preceding ones, are those of the Orinoco tar belt (Faja bituminosa) in Venezuela. Deeper than Athabasca tar sands, about 600 to 1000 m, less altered by surface water circulation, and located in a warmer country, their average temperature is about 50 °C and their viscosity is less than 10^4 cP, which does not therefore classify them as bitumens.

Due to their frequent presence on the surface or at shallow depths, bitumen and extra-heavy oils have been known for a very long time. Bitumens were used in antiquity for caulking boats, joints in construction, lighting, and often even as medicine. The extra-heavy oils of the Pechelbronn deposit (Pechelbronn means source of pitch, ancient name of the bitumen) in Northern Alsace were used since the Middle Ages. The Athabasca deposits began to attract attention at the beginning of the 18th century. Exploited from the 1960s, it was not until 2000 that this exploitation really grew, in particular thanks to the very high increase of the price of the oil, which took place from 2003 to 2014. The extra-heavy oils of Venezuela began to be inventoried in the 1930s and their start-up began in the 1960s, but as in Athabasca, it was not until 2000 that they became fully exploited.

Many molecules are found in petroleum, the chemical structure of which is evidently parent to that of molecules that constitute living organisms. These molecules are called geochemical fossils or more often biomarkers. Among these, porphyrins, the structure of which is very similar to that of the tetrapyrrole nucleus of chlorophyll (see Appendix 1), was first described in 1934 by the German chemist Alfred Treibs, the first to have thus provided chemical evidence of the organic origin of oils. Since then, thousands of these biomarkers have been identified, following the amazing progress of analytical methods from the 1970s onwards, and more particularly the use of gas chromatography coupled to mass spectrometry (GC-MS).

1.2 The natural gases

They are also made up largely of hydrocarbons. The hydrocarbons contained in natural gases are in principle those, which are in a gaseous state under the so-called normal conditions (15 °C, atmospheric pressure). They are those whose number of carbons is less than or equal to 4: the most abundant is methane, CH_4, followed by ethane, C_2H_6, propane, C_3H_8, then butane and isobutane, C_4H_{10}. Natural gases also contain, in some cases very large amounts, non-hydrocarbon gases, the main ones being carbon dioxide, CO_2, nitrogen, N_2, and hydrogen sulphide, H_2S. Sometimes there are a few rare gases, helium and argon, or even hydrogen, H_2 (Table 2). Some fields also contain traces of arsenic and mercury[3], which must be eliminated. If hydrocarbons in gases have an organic origin, non-hydrocarbons may have in part a mineral origin, or even an organo-mineral origin, by reaction of the hydrocarbons on the reservoir rocks of their deposit *(Goldstein and Aizenshtat, 1994)*.

Here again, there is not one but natural gases, and their variety is very large (Table 2).

This diversity in the composition of natural gases results, among other things, in a variety of physical properties and in particular their calorific value, which is the source of their economic interest (see Chapter 2.2). It is well understood that the discovery of a field rich in nitrogen or carbon dioxide has much less interest than that of an almost pure methane field.

But in the subsoil what is called a natural gas field is in fact a fluid phase whose main constituents are generally under conditions of temperature and pressure called supercritical. This fluid phase has physical properties close to those of a gas. If it consists essentially of molecules which would be gaseous at the surface, it also contains, in the dissolved state, sometimes in a high proportion, liquid hydrocarbons

---

[3] Mercury in particular can corrode gas production and processing facilities, and thus has been the cause of serious accidents.

Table 2	Variation of the proportions, as a percentage of the total volume, of the main constituents of natural gas fields, after Sokolov (1974). CO_2 and H_2S originate in the reaction of hydrocarbons on carbonates and sulphates contained in reservoir rocks, and helium (He) is formed from the decay of radioactive material. Nitrogen is formed from the release of nitrogen by the thermal cracking of kerogen (see chapters 2-3). All these components are mostly found in ancient and/or once deep gas reservoirs.

Fields	Age	CH_4	C_2-C_6	CO_2	N_2	H_2	H_2S	He
Kane (United States)	Cenozoic	99.3	0.4		0.1	0.1		0.15
Sweetwater (United States)	Carboniferous	75.6	1.3	2.7	20.2			0.75
Newago (United States)	Carboniferous	85.5	1.6	0.4	12.4			1.1
Spence (United States)	Carboniferous	91	4.8	0.1	3.6			0.14
Transylvania (Romania)	Miocene	98–99	0.8	0.5	1–2			
Krecsegopan (Hungary)	Miocene	42–47		45–83	3–6	0.9		
Schlochteren (The Netherlands)	Permian	81.3	3.5	0.8	14.4			
Lacq (France)	Jurassic	74	2	9			15	
Northern Germany Plain	Carboniferous-Permian	Up to 95	0.3–12	Up to 60	Up to 99	Up to 70	0–8	
Cumato, Sakada (Japan)	Quaternary	42–98	0.1	0.5–4.5	4–53			
East England	Carboniferous-Permian	90			10			
Weedhorff (Germany)	Carboniferous-Permian	92.5	1.7	1.6	4.2			
Baden (Germany)	Carboniferous-Permian	82.1	0.8	10.3	6.8			
Touïmazi (Volga-Urals, CIS)	Devonian	39.5	49.8	0.1	10.6			
Sokovka (Sub-Caucasus, CIS)	Cretaceous	76	17	5	1			
Plain of the Pô (Italy)	Pliocene	99	1					
Plain of the Pô (Italy)	Miocene	90	Up to 8					
Hassi R'Mel (Algeria)	Trias	80	15		5			
Piaui (Myanmar)	Trias	88.1	2.6	0.7	8			0.003
Ourengoy (West Siberia)	Cretaceous	98.5	0.1	0.2	1.1			0.01

called C$_5$ + hydrocarbons (i.e. a number of carbons greater than 4), which would therefore be liquid under surface conditions. This phase "gas" has a density relative to water, which can vary from 0.1 to 0.4 depending on its composition, as well as the temperature and the pressure prevailing in the deposit.

It may constitute the deposit alone or form part of a mixed oil/gas deposit in which it surmounts an oil (petroleum) phase of higher density, mainly formed of liquid hydrocarbons, but which may also contain dissolved molecules that would be gaseous on the surface. The gas dissolved in the oil phase is referred to as the associated gas: the gas oil ratio (GOR), expressed in m^3 of gas under normal conditions per m^3 of oil, is characteristic of the quantity of "gas" dissolved in this liquid phase. The "gas" phase surmounting the oil phase is called a "gas cap".

During production of the deposits, liquid hydrocarbons, which are present in the dissolved state in the gas phase, condense in part. When they rise towards the surface, this condensation occurs in the deposit or in the wells, but especially at the wellhead, in a series of separators in which the initial pressure is gradually lowered. The liquid hydrocarbons thus recovered are so-called condensates (C). The gas then goes to a processing plant, where remaining water, minerals and non-hydrocarbon gases are removed. The so-called natural gas plant liquids (NGPL) are then recovered by refrigeration. They are C$_5$ + hydrocarbons which have not been condensed at the well-head, and hydrocarbons which are gaseous under normal conditions other than methane: ethane, propane and butanes. Propane and butanes will be liquefied under pressure to form liquefied petroleum gas (LPG). Ethane is partly found in clean (dry) gas and partly in LPG.

The sum of condensates (C) and natural gas plants liquids (NGPL) is called natural gas liquids (NGL).

The gas at the outlet of the well is called gross gas. In international accounts, condensates and NGPLs extracted from this gross gas are accounted for in the oil category, while the remaining clean gas called dry gas is included in the gas category.

These differences between gas in the form of a gas phase in a reservoir with or without an oil phase, and gas in the form of gaseous molecules recovered at the surface can be better understood by examining Figure 4. Six situations are considered, by increasing importance of the gas phase in the field. We shall return to this question in the second part.

Gas processing plants may be located on gas fields. Dry gas is evacuated by pipeline and LPG by specialized trucks. But often, and necessarily if the deposit is at sea, the unprocessed or partially processed gas is shipped by gas pipeline to LPG extraction terminals located at the seaside. LPG, and dry gas after liquefaction at –161 °C (liquefied natural gas, LNG) may be transported by specialized vessels. In the case of LNG, LNG tankers will be transported to regasification facilities located on the seafront in the consuming countries.

If the gas is associated with an oil field but does not economically justify the construction of gas transportation means the gross gas is re-injected into the deposit or flared.

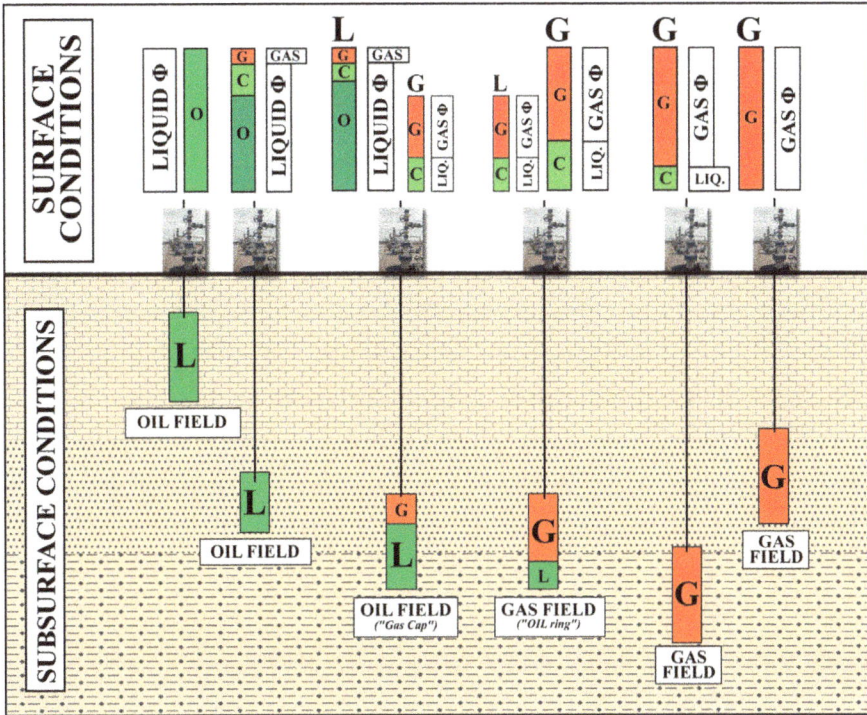

Figure 4 *Relationship between oil phase and gas phase in a field (bottom conditions) and in surface conditions: L: liquid phase; P: petroleum; C: condensates; G: gas phase. As a function of the proportion of gaseous molecules in the surface conditions in the whole of the deposit, 6 situations are represented from no gas and condensate at all to only gas. The gas recovered at the wellhead still contains liquid hydrocarbons that will be recovered along with propane and butane in natural gas processing plants in the form of natural gas processing plant liquids (NGPL). Courtesy Roland Vially.*

These natural gases have essentially the same way of formation as petroleum, that is to say the thermal degradation of organic matter contained in certain sediments, as will be seen later. They are very frequently associated with oil in the same geological structure, as described above. This is called thermal gas.

The shale gases (in fact source-rock gases) which are at the moment in the news are thermal gases. They have the peculiarity, like shale oils (source-rock oils), of being located in extremely low permeability reservoirs located in or in contact with the rocks that gave birth to them, their source-rocks (see Chapters 2.3 and 4). They have been known for a long time, since the small town of Fredonia in New York State was fuelled with gas of this nature as early as 1840. Technological advances now make it possible to exploit it on a large scale in the United States (see second part).

But there are also, which is not the case for petroleum, fields of biogenic gas (bacterial, biochemical), which come from the action of methanogenic bacteria on organic substances dissolved in the water of specific sedimentation media *(Vially et al., 1992)*.

This is also how the marsh gas is formed, or the biogas produced by agricultural fermenters.

Biogenic gas, unless mixed with a gas of thermal origin, contains no hydrocarbon other than methane.

There is also methane of mineral origin formed by the action of water on minerals at temperatures of a few hundred degrees in the volcanic zones of the earth's crust. But it does not form exploitable deposits.

1.3 The Coals

These are sedimentary rocks containing a very high proportion of organic carbon from plant debris. Formally, the name coal is reserved for rocks containing more than 40% of their dry weight in organic carbon. But it will be seen that the coals are associated, in what are called the coal series, with rocks containing the same type of organic matter, but in a lesser content.

Coals can come from the accumulation of unicellular algae debris, which are so-called inferior plants. These are algal coals (algal coals, sapropelic coals, bogheads). They can also come from accumulations of spores and pollens. These are spore coals (cannel coals). But these are anecdotal categories, because in their overwhelming majority coals come from the accumulation of debris of so-called higher plants, trees and herbaceous plants, which are rich in lignocellulosic tissues (wood, leaf veins, roots, etc.) It is then humic coals, or coals *sensu stricto*.

Coals are not, as we sometimes read, solid-state petroleum, as are bitumens, or solid hydrocarbons, although, as we shall see, they can give birth to oil and sometimes even still contain some. They are very complex organic mineral rocks, the organic constituents of which are mostly derived from constituents of the higher plants. In optical microscopy by reflection, we observe what are called macerals[4] by analogy with the minerals of the common rocks.

The main macerals are:

- **vitrinites**, 70% on average of the surface observed under the microscope, are solid amorphous substances (solid gels): they form in sediments from lignin, cellulose and hemicellulose, which are the main constituents of higher plants support tissues;

---

[4] The term maceral was created in 1935 by the paleobotanist and famous Scottish feminist Mary Stopes (1880–1958).

- **liptinites**, which include the **exinites** formed from the "external" constituents of the higher plants: sporinite from spores and pollens, cutinite from leaf cuticles, resinite from tree resins, etc. and the **alginites**, which is formed not from higher plants, but from unicellular microscopic algae (lower plants) living in the sedimentation media;

- **inertinites**, the constituents of them are most often derived from the above macerals by oxidative phenomena, forest fires or oxidation during transport to the sedimentation medium.

These major categories are distinguished in particular by their reflectance and fluorescence (Figure 5).

The total solid organic fraction of coals, as for all sedimentary rocks containing debris and other organic substances issued of organisms or from their degradation after death, is called kerogen. Kerogen is in practice defined in terms of its insolubility in the usual organic solvents (*Durand et al., 1980*). Macerals are petrographic categories: it is the parts of this kerogen that under the microscope are visible and whose origin is identifiable. Their identification is the subject of so-called coal petrography, one of the most important works of reference being that of *Stach et al. (1982)*.

The mineral fraction of the coals is in the form of fine individual beds, in particular clays, and constituents intimately mixed with the macerals, clays, quartz, but also iron sulphide (pyrite) or zinc sulphide (sphalerite, blende) containing traces of many other metals than iron and zinc[5]. There are also organo-mineral compounds, undetectable under the microscope.

Coals, like any sedimentary rock, also contain water, about 50% to 5% depending on their nature and geological history (see below).

When burning a coal, the whole mineral fraction produces ashes. These represent on average about 15% of the dry weight of the initial coal.

1.4   The bituminous shales (oil shales)

They are, like coals, sedimentary rocks exceptionally rich in organic carbon, but on average much less than coal. The usual contents are of the order of 10–15% by weight of organic carbon relative to the dry weight of the rock. In addition,

---

[5] Coals often have higher contents than average contents of sedimentary rocks (average contents are called Clarke in memory of the American geochemist who first established these values for the earth's crust) in a number of elements, often present in traces in sulphides such as pyrite and sphalerite. Among those, antimony, arsenic, beryllium, cadmium, chromium, cobalt, fluorine, germanium, lead, manganese, mercury, nickel, selenium, thallium and vanadium are considered to be harmful at low doses to health. Coals also frequently contain sulfur, present in metallic sulphides but also in organic compounds. Combustion of this sulfur produces sulfur oxides, which are also harmful. There are also variable but sometimes significant quantities of radioactive elements, uranium and thorium and their descendants.

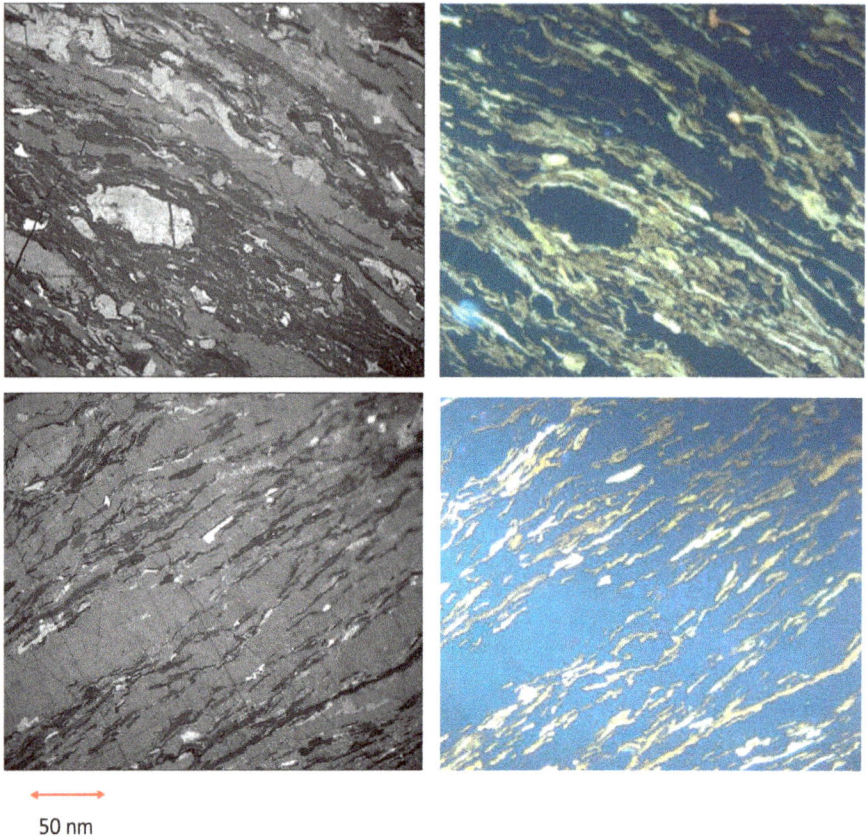

50 nm

Figure 5 *Micrographs of two polished sections of a coal taken from a well at Gironville in Lorraine, France, examined by reflection under an immersion oil objective (refractive index of oil 1.518): on the left in monochromatic visible light (wavelength 546 nm) and on the right in monochromatic UV light (wavelength 365 nm). They show the great heterogeneity of the coals: vitrinite is seen on the left as gray areas encompassing more reflective inertinite inclusions, and beds of liptinite which are very little reflective, therefore black. Liptinite (mainly sporinite derived from plant spores and cutinite derived from cuticles of leaves, whose outlines are visible) is black in visible light and fluoresces in yellow under UV light, while non-fluorescent vitrinite and inertinite are black. At the bottom right, the section has a general blue tint. This is due to the dissolving by the immersion oil of some hydrocarbons present initially in the coal sample (see Chapter 3.2.1), which have a blue fluorescence. Coal petrography is widely used to determine what is called the coal rank (see Chapter 3.2.1), thanks in particular to the measurement of the reflectance of vitrinite. A detailed description of these methods and their use can be found, for example, in Stach et al. (1982). Courtesy: Professor Luis Martinez, University of Strasbourg and École and Observatoire des Sciences de la Terre.*

the solid organic fraction (kerogen) is not derived from higher plants, but mainly from lower plants lacking lignocellulosic tissues: unicellular algae debris and bacterial membranes, in which reflection microscopy cannot make distinctions because it is presented in amorphous masses with very low reflectance power and fluorescence color too uniform. A small part of these debris is however sometimes identifiable by transmission microscopy.

Added to this is an additional requirement: the ability for a bituminous shale (also often called oil shale) to produce in an economically cost-effective manner a synthetic oil (synfuel) known as shale oil[6], by thermal treatment in the absence of air (pyrolysis) at about 500 °C. We shall see that it therefore must never have reached great depths during its geological history. Moreover, for economic reasons, their exploitation can only be considered if they are on the surface or close to the surface.

The terms bituminous shale and oil shale are therefore unfortunate, although they are consecrated by usage. These rocks do not contain bitumen, which, as we have seen, is a solid variety of petroleum, but kerogen, which is, as we shall see, constituted by the remains of organisms which have lived in the sedimentary medium or brought from the neighboring continent. They do not either contain oil, their burial having not been sufficiently important for the formation of petroleum.

Bituminous shales (oil shales) are also frequently confused with the bituminous (tar) sands, which are, as we shall see, sandstones containing bitumen, the residue of the alteration of an oil deposit.

---

[6] The term "shale oil" is also used to refer to oil that has remained trapped in the sediment that gave birth to it, its source-rock. The latter, in general, has a weaker kerogen content than a bituminous shale, and has, as we shall see, been deeply buried for producing oil. This term has therefore become a source of confusion. To dispel it, this source-rock oil is now called light tight oil (LTO) rather than shale oil. But the term oil shale is very often used in an inappropriate manner to designate the source-rock that contains the LTO. The term kerogen shale, however, begins to appear to designate true oil shale (bituminous shale), and that of kerogen oil to designate the oil produced during its pyrolysis. These problems of semantics will be better understood in the rest of this book, once the mechanisms of oil and gas formation have been explained.

2

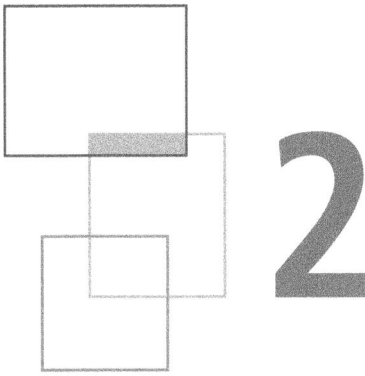

How do fossil fuels form?

2.1 Their origin: kerogens, debris from organisms accumulated in certain sediments

Fossil fuels form from sediments that contain, together with minerals, microscopic debris, more or less altered, of organisms that have lived in the sedimentary medium, or debris brought from neighboring land. These organisms were mostly lower plants, marine or lake phytoplankton, and higher terrestrial plants (trees and herbaceous plants). Various organic residues, such as fecal pellets, exudates, vegetable resins, soot from forest fires and soils, or organic molecules adsorbed on mineral surfaces, are also present.

The solid part, more precisely the part insoluble in common organic solvents, of these debris and organic substances is called kerogen (*Durand et al., 1980*). Kerogen is therefore a chemical category. Its elementary analysis reveals that it consists essentially of 5 elements: in decreasing order of their average quantity, expressed in proportion to the total number of atoms present, hydrogen, carbon, oxygen, and some sulfur and nitrogen.

Kerogen containing sediments are found among other sediments in sedimentary basins: these are stacks of sediment strata accumulated over geological time. Their dimensions, surface and depth, are very variable, and their geometry can be extremely complex. A relatively simple example is the Paris basin, the main French sedimentary basin, with a surface area of about 150,000 km² and a current maximum depth of about 3 km. It began to form about 250 million years ago, at the beginning of the Mesozoic era, on a Paleozoic basement.

The keys to the accumulation of kerogen in a sediment are on the one hand the abundance of inputs of organic matter during its deposition and on the other hand what is called anoxia, i.e. the absence of oxygen, or at least a limited presence thereof, in the water of the sedimentation medium and of the sediment (*Tissot and Welte, 1978; Demaison and Moore, 1980; Bordenave et al., 1993*).

The normal destiny of biological molecules after the death of organisms is indeed to be recycled by other organisms that feed on them. In the presence of oxygen, the so-called aerobic organisms and microorganisms, i.e. whose metabolism is based on oxygen, degrade these molecules very quickly to carbon dioxide and water essentially. In the absence of oxygen, these aerobic organisms cannot survive, and despite the presence of other microorganisms, anaerobic this time, degradation of organic matter is much slower.

The formation of kerogen thus takes place thanks to a "leak" of the biological cycle of organic carbon, leak which feeds the sedimentary cycle of the latter. The proportion of organic carbon in the biomass that escapes the biological cycle is very low and impossible to assess very precisely. The most common estimates are about 0.1% by mass of the organic carbon of the biomass synthesized each year in the world, which is thus incorporated into sediments. But the immense lengths, tens to hundreds of million years, of the sedimentary cycle mean that there are about 10,000 to 100,000 times more organic carbon stored in kerogen containing sediments than in the annual biomass production.

Anoxia is more easily carried out in very fine-grained sediments, clays and marls, because they are very low permeability sediments, where the water and the dissolved oxygen that it may contain circulate with great difficulty. It is in this type of rocks that we find most of the kerogens. However, the palaeogeographic situation of the deposit environment also has its importance: deposits in shallow calm waters of lagoons, bays or seas protected by thresholds or bottlenecks of oxygen inputs by ocean circulation, or lakes, favor the accumulation of kerogens: productivity in planktonic organisms is often important. On the other hand, the transport distance between areas of high biological productivity and sediment is low, which limits oxidative degradation before incorporation into sediments. Areas of upwelling are also favored, i.e. coastal areas where water flows from the depths, as is currently the case on the western coasts of Chile and Mauritania. Mineral salts dissolved in the water coming from the depths results in an exceptional planktonic productivity in the euphotic zone, that is to say the slice of water where sunlight penetrates. Anoxia is created here by the consumption of the oxygen available in the water under the euphotic zone by aerobic microorganisms, which results in the destruction of only a part of the organic debris in the process of sedimentation, the remainder being thus preserved.

In the swamps and deltas of rivers of climatic zones favorable to plants productivity, in clay sediments accumulate in places debris of so called higher plants (trees, herbaceous plants). Thus, when these accumulations are particularly important, peats[7], which eventually give birth to coals, are formed. Some of these plants debris is also exported to the sea.

There is therefore not one but a variety of kerogens, depending on the origin of the initial organic debris and the sedimentation conditions.

The organic matter which the kerogens originate in is almost essentially vegetal: indeed, the world's biomass is about 90% made up of plants, roughly equal parts of marine or lake plants (mainly phytoplankton which consists mostly of unicellular organisms: microalgae and photosynthetic bacteria), and terrestrial higher plants (trees, herbaceous plants) (*Huc, 1980*).

There are three reference types of kerogen (*Durand, 1980*) characterized by typical series: those derived from a lacustrine planktonic biomass (type 1), those derived from a marine plankton biomass (type 2), and those derived from terrestrial higher plants (type 3). The most frequent are types 2 and 3, type 1 having an anecdotal character on a global scale. There are intermediates, depending on the variety of marine and terrestrial inputs, and the conditions under which they were transported prior to their incorporation into the sediment.

As already said, the most abundant elements of the kerogens, in proportion to the number of atoms present, are hydrogen, carbon, and oxygen[8], and then sulfur and nitrogen. Their initial contents (we will see that they change during the burial of the sediments) vary according to the kerogens. Types 1 and 2, which are mainly derived from planktonic biomass, are richer in hydrogen and less oxygen rich than type 3, which is derived mainly from a biomass of higher terrestrial plants.

Type 2 kerogens are often more sulfur-rich than average, especially when the sedimentation medium is low in dissolved iron, which is the case for sedimentary series rich in carbonates. This is due to their formation in the marine environment: in the sediment where organic debris accumulates, so-called sulfate-reducing bacteria produce hydrogen sulphide from sulphates dissolved in seawater. When iron is present, this hydrogen sulphide combines with it to produce iron sulphide (pyrite). When iron is not present, hydrogen sulphide combines to kerogen to produce sulfur compounds instead of forming pyrite from iron. Sulfur compounds will later be partly found in the oils formed from this type of kerogen.

Sediment concentrations in kerogens of marine (type 2) or lacustrine (type 1) origin are extremely variable, but an organic carbon content of more than 2% of the dry weight is not frequent and a content of more than 5%, as it may be the case for bituminous shales, exceptional (Table 3). The so-called source-rocks, i.e. the kerogen

---

[7] The peats from which coals derive are formed rather in hot climates and are subjected to a rapid burial. They originate from plant associations different of the mosses and sphagnums well-known in cold swamps, which generally do not give birth to coals because they are barely sufficiently buried (see further on).

[8] One might think that, being formed in the absence of oxygen, the kerogens do not contain it. In fact, their oxygen comes from the biological molecules from which they are derived.

Table 3 *Table of Frequencies of Organic and Mineral Carbon Content, in percentage dry mass, of a set of over 10,000 samples of sedimentary rocks from oil basins analyzed at the Institut du pétrole et des énergies nouvelles (IFPEN). Each cell represents the percentage of samples whose organic and inorganic carbon contents are within the limits indicated on the abscissa and on the ordinate. The mineral carbon content increases from 0 to 13%, which is the maximum possible (for comparison, the pure calcium carbonate contains 12%), and the organic carbon contents here are arbitrarily limited to 5%. Samples with less than 5% organic carbon therefore account for about 93% of this sampling, and those with less than 1% of organic carbon about 60%. It is observed that statistically, the organic carbon content is lower the greater the mineral carbon content.*

% Organic carbon

% Mineral Carbon	0-0,5	0,5-1	1-1,5	1,5-2	2-2,5	2,5-3	3-3,5	3,5-4	4-4,5	4,5-5	Total
0-1	8,2	5,2	3,8	2,1	1,5	0,7	1	0,8	0,4	0,4	24,1
1-2	4,4	2,9	1,8	1,7	1,2	0,9	0,5	0,6	0,2	0,2	14,4
2-3	2,9	1,9	1	0,8	0,5	0,3	0,1	0,1	0,1	0,2	7,9
3-4	2,4	2,2	1,3	0,6	0,3	0,3	0,1	0,1	0,1	0,2	7,6
4-5	2,4	2,6	1	0,5	0,2	0,2	0,2	0,2	0,1	0,1	7,5
5-6	2,4	1,6	0,7	0,4	0,3	0,1	0,2	0,1	0,1	0,1	6
6-7	1,9	1	0,5	0,3	0,2	0,2	0,2	0,1	0,1	0,1	4,6
7-8	2,5	1,4	0,5	0,3	0,4	0,2	0	0,1	0	0,1	5,5
8-9	2,3	0,8	0,3	0,2	0,1	0,1	0,1	0	0	0	3,9
9-10	2,2	0,6	0,3	0,2	0,1	0	0	0	0	0	3,4
10-11	2,9	0,7	0,2	0	0	0	0	0	0	0	3,8
11-13	3,9	0,6	0	0	0	0	0	0	0	0	4,5
Total	38,4	21,5	11,4	7,1	4,8	3	2,4	2,1	1,1	1,4	93,2

containing sediments, which gave birth to oil and gas, had mostly only a few percent organic carbon contents at the time of their deposition.

Kerogen is found in these rocks as a discontinuous network of particles, clusters or laminae, sometimes observable to the naked eye for particularly rich rocks (Figure 6a), but which are mostly well visible only by special microscopy techniques (Figure 6b),

Kerogens of terrestrial higher plants origin (type 3) also have very variable concentrations in their sediments, but sometimes they are concentrated in very rich veins, metrically thick and extended in kilometers, which will give birth to the exploitable veins of coal.

Some geological periods have been much more favorable than others to the accumulation of kerogens in sediments. For kerogens of marine plankton origin (type 2), they are mainly Silurian (around 430 million years ago), the end of the Devonian and the beginning of the Carboniferous (around 360 million years ago), the end of the Jurassic (around 160 million years ago) and the middle of Cretaceous (around 100 million years ago).

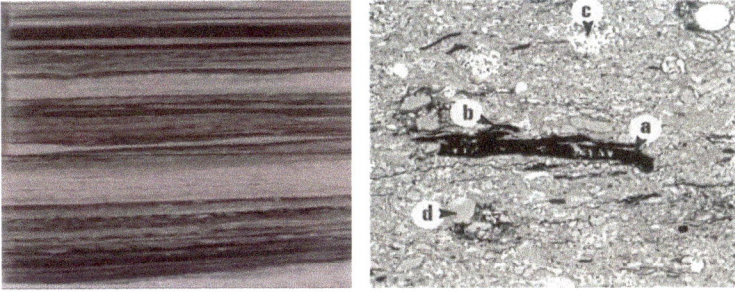

Figure 6 *(left) Dark-colored kerogen-rich laminae, which can be observed with the naked eye on a core of about 5 centimeters diameter drilled in the Kimmeridge Clay Formation, the main source-rock of the North Sea. The thicknesses and the kerogen content of the laminaes are very variable, reflecting the variation in the intensity of the inputs of organic matter and the influence of the sedimentation conditions. (right) Laminaes and kerogen clusters, seen here in dark black on a polished section with a backscattered electron scanning electron microscopy technique, forming a discontinuous network in a predominantly clayey rock. The bulkier particle (a) is probably a plant debris. Its length is approximately 60 µm. b is a mixture of kerogen and clay, c is pyrite, d is detritic quartz. The rock is also a sample of Kimmeridge Clay. The organic carbon content of this sample is about 3% of the dry weight. Courtesy Compagnie Total and Claude Laffont (IFP School) for Figure 6a, and Belin (1992) for Figure 6b.*

Figure 7 *An argillite (sediment made of predominantly clay minerals) very rich in kerogen (about 10% dry weight in organic carbon). This rock, which is a sample of the Schistes-Cartons Formation of the Lower Toarcian of Lorraine, was used on its outcrops as a bituminous shale until shortly after the Second World War). Courtesy Roland Vially.*

37

These exceptional accumulations are generally linked (*Huc, 2005*) to fluctuations in the level of the oceans, an increasing level (marine transgression) increasing the surface of the marine domain and in particular shallow seas, conducive to this accumulation, on the periphery of the continents. For kerogens of terrestrial plant origin (type 3), their accumulation would, on the contrary, have been favored by periods of declining ocean level (marine regression), favoring the development of the continental sedimentation domain at the end of the Carboniferous period and the beginning of the Permian (around –300 million years ago) but also during the Oligocene-Miocene (about –20 million years ago).

2.2 Their sites of formation: sedimentary basins, depressions of the earth's crust invaded by the waters

During geological times, generally under the effect of tectonic forces working in extension, depressions are formed in the Earth's crust. These are gradually invaded by waters to form oceans, seas and lakes. These are the sedimentary basins, of which the classic example is in France the Basin of Paris. There are deposited minerals produced in an aquatic environment: carbonates in the form of mineral parts (tests) of microfossils or debris of shells, evaporites (chemical precipitates resulting from the evaporation of seawater) such as salt, potash, carbonates and sulphates. There are also mineral constructions of biological origin such as coral reefs and oolites, and debris of microorganisms living in aquatic environments (plankton, bacteria). Other constituents include sand and clay minerals, and organic debris (plant debris, organic matter from soils...), which come from the neighboring land, brought by rivers and by the wind. These sediments accumulate by strata, depending on the fluctuation of water levels (at the small-time scale under the influence of climatic variations and on a larger time scale of tectonic phenomena) and distances to the coast, during the deepening of the basement of the basin. This deepening is called subsidence: this is caused by tectonic forces, but also by the weight of accumulated sediments. The sediments, formed in an aquatic environment, have their porosity saturated with water, and expel it gradually during their burial, according to their mechanical characteristics and their permeability. Their volume thus decreases, this is called compaction, and their density increases.

The dimensions of the sedimentary basins are very variable, and often considerable: the Paris Basin (Figure 8) has an area of about 150,000 km^2 and a maximum current depth of about 3000 meters. The Western Siberia Basin covers an area about ten times larger, and the present depth of the Southern Basin of the Caspian Sea is more than 20 km.

Deepening can be interrupted, and when the tectonic forces work this time in compression, the basement of the basin and the sediments that it supports can rise towards

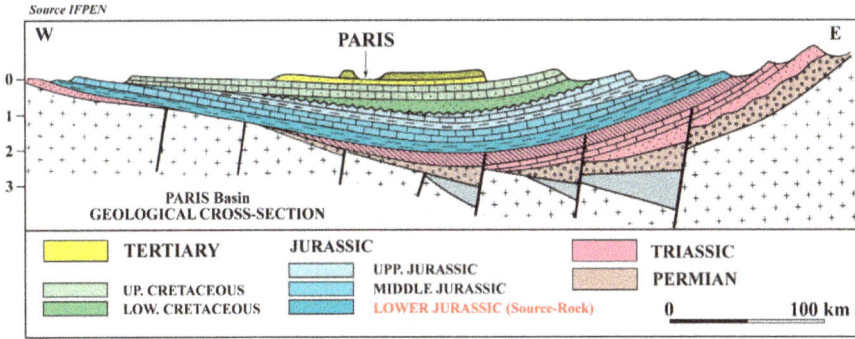

Source IFPEN

W   PARIS   E

PARIS Basin
GEOLOGICAL CROSS-SECTION

TERTIARY	JURASSIC		TRIASSIC
UP. CRETACEOUS	UPP. JURASSIC		PERMIAN
LOW. CRETACEOUS	MIDDLE JURASSIC		
	LOWER JURASSIC (Source-Rock)	0	100 km

Figure 8   *Schematic section of the Paris Basin from the Vosges to Normandy. It is a stack of sedimentary strata of the Mesozoic (Secondary Era) and Cenozoic (Tertiary Era) ages, on a fractured basement of the Paleozoic age (Primary Era) containing small Carboniferous age basins (in grey colour). The colors of the strata according to their age are the conventional colors in geology. Small oil fields have been found in the Triassic and the Jurassic strata. The source-rocks of these pools, that is to say the rocks where their oil has formed, are found in the Lias (Lower Jurassic). Courtesy Roland Vially.*

the surface. This is called a tectonic inversion. The basin then emerges and no longer receives sediments. It becomes a fossil basin. The accumulated sediments are eroded as they rise and are redistributed in the active sedimentary basins of the vicinity. The Paris Basin (Figure 8), created 250 million years ago, is now in this situation as a result of a tectonic inversion that began about 80 million years ago. The uplift and subsequent erosion are most pronounced in its southeastern part. They result mainly from the forces of compression, which caused the formation of the Alps, due to the collision between the Eurasian Plate and the African Plate.

The stacking of strata, the so-called sedimentary series, generally contains very little sediment rich in kerogens. In the Basin de Paris, these sediments are mostly located in the lower Jurassic (Hettangian-Sinemurian and Lower Toarcian, from −200 to −180 million years, in dark blue in Figure 8). At that time there was a shallow sea in this place, which communicated little with the world ocean, but it was very extensive (it also covered northern Germany), and the conditions for high planktonic productivity were met.

2.3   The key to their formation: the thermal history of kerogen containing sediments

The terrestrial globe is hot, and its heat has various origins: the remaining heat of its formation, the radioactivity of the rocks it contains (it is the most important source, some say 80% of the heat produced), the change of liquid phase into solid phase at the periphery of the core, mechanical friction. This heat is evacuated to the space and the corresponding heat flux is in total about 42 TW[9], i.e. an average flux of 82 mW per m^2 of the Earth's surface. This terrestrial flux can vary significantly from one place to another, but also over time, depending on local tectonics and its evolution, but it remains very low in comparison to the average flux of solar radiation absorbed by the Earth's surface, which is about 170 W/m^2 or 2000 times more than the average terrestrial flux. However, this flow of solar energy being returned in different ways to space, affects the temperature of the terrestrial globe, via the absorption by the ground of the solar radiation and the greenhouse effect, only down to a depth of about 100 meters. It is therefore the flow of heat of deep origin that essentially governs the increase in the temperature of the rocks below a few tens of meters. It is responsible in particular for the convection movements in the mantle, the drift of the continents and, as we shall see, the formation of fossil fuels.

Between the core, whose temperature is of the order of 6000 °C and the surface, whose mean temperature, controlled by the greenhouse effect, is on average of the order of 15 °C, there exists a temperature gradient, which at the crossing of the sedimentary basins strata is on average of the order of 30 °C/km. However, variations can be very important in a range of 10 to 80 °C/km, depending on the local geothermal flux, the thermal conductivity of the rocks traversed, local heat sources (radioactive rocks, volcanic magmas)[10], or the importance of water circulation in faults and aquifers.

A sedimentary rock thus sees during its burial its temperature increase at a rate of 30 °C per km on average. The rocks at the bottom of the Paris Basin, at a depth of 3 km, have, for example, currently a temperature of about 100 °C (30 °C/km × 3 + 10 °C of average ground temperature). Conversely, its temperature will decrease if it rises during a tectonic inversion.

The corresponding heating rate is extremely low on a human scale. The maximum temperature reached at the bottom of the Paris Basin, about 150 °C, was about 150 million years ago, which corresponds to an average heating rate of about 1 °C per million years! The mean rate of burial of the initial sediments was of the order of 2 mm per century!

Geologists are now able to set up with a fairly good approximation the curves of the temperature evolution of the rocks they contain as a function of geological time, that is to say their thermal history (Figure 9).

A synthetic mode of representation which combines the history of burial and thermal history is that of Figure 10.

---

[9] The watt (W), unit of power of the International System of Units (SI) corresponds to an energy quantity of 1 joule (J) mobilized per second.

[10] Unusually large gradients, up to several hundred °C per km, may exist locally in the vicinity of magmas in active volcanic areas. This is the case, for example, in Larderello, Italy, where heat is used to produce electricity.

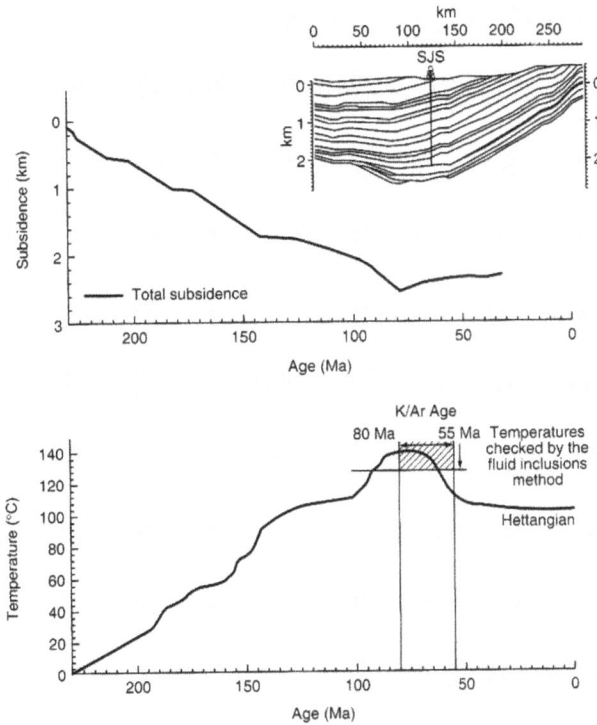

Figure 9 *Subsidence and thermal histories of the Hettangian strata crossed in the Saint-Just-Sauvage well (SJS), located on the simplified section of the Paris Basin in up and to the right of the Figure. A fairly regular burial and temperature increase is observed up to about 80 million years ago, followed by a tectonic inversion causing a rise and a cooling of the strata. The maximum temperature reached by the Hettangian at this location, determined by the so-called fluid inclusions method, was about 140 °C. Durand, 2003 according to Burrus, 1997, Courtesy of Editions Technip.*

The sedimentary basins can thus be considered as immense physicochemical reactors loaded very slowly from the top, deepening, filling and gradually deforming, until a tectonic inversion occurs which causes the uplift of their contents and erosion.

As low as they are, the temperatures and heating rates are sufficient to cause, at the enormous time scale of geological eras, considerable changes in the rocks and, in particular, the formation of fossil fuels in sediments containing kerogen.

When a sediment containing kerogen buries, the latter is subjected to very slowly increasing temperatures and is gradually baked as in an oven, but in the absence of oxygen since there is none available in the free state in the sediments.

Cooking in the absence of oxygen is what is known in scientific terms as pyrolysis or thermal cracking. It causes a breakdown of large organic molecules into smaller

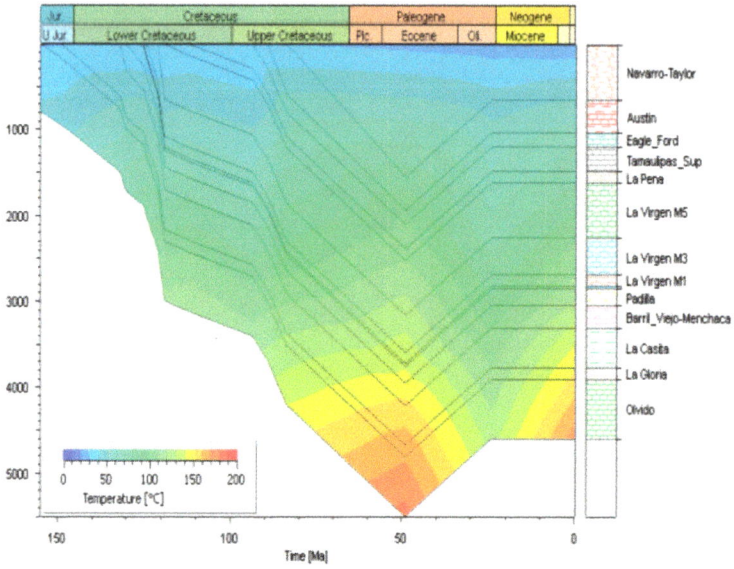

Figure 10	Synthetic reconstruction of burial histories and thermal histories for the different strata of the sedimentary series of a Mexican basin. In this representation, isotherms are identified by colored codes on the canvas constituted by the burying curves of the boundaries of the different sedimentary formations (indicated here with their name, their mineralogy and their current thickness). This allows visualizing both the history of burial and thermal history for each sedimentary stratum. In this case, after a fairly regular burial of the series for more than 100 million years, a tectonic inversion begins about 50 million years ago, which continues until about 25 million years ago. First, this inversion is accompanied by a cooling of the strata. Then the depths of the strata stabilize, and their warming is observed, probably under the influence of an increase in the geothermal flux. At the base of the deepest stratum, the maximum temperature reached was at 5500 meters of about 190 °C, 50 million years ago. Be careful not to confuse the colors of geological ages of the legend above, which are conventional colors, with the colors indicating the temperatures on the graph! Courtesy: Professor Luis Martinez, University of Strasbourg and École and Observatoire des Sciences de la Terre.

and smaller molecules. The familiar image of the cooking makes it easy to memorize the succession of events that takes place as burial increases. If a roast (kerogen) is cooked in a well-closed casserole without oxygen, it first produces water and carbon dioxide, then organic juices, then gases, and the solid part of the roast is transformed progressively in an increasingly carbon-rich residue. On the other hand, the cook knows from experience that this succession of phenomena occurs all the more rapidly as the temperature is higher in the casserole.

The same is true for kerogen in sediments (Figure 11). However, temperature plays very little in the first stage of its degradation, called early diagenesis. This is mainly

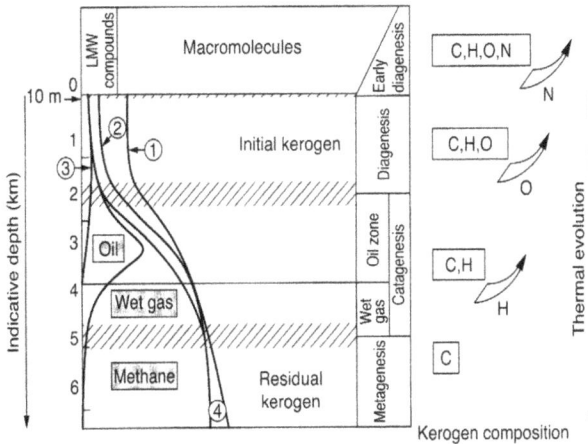

Flowsheet of the formation of oil and gas from kerogen as a function of depth: 1: low molecular weight oxygenated compounds (CO_2, H_2O...); 2: resins and asphaltenes; 3: biomarkers; 4: insoluble pyrobitumen (coke) formed from oil as a consequence of gas formation. After Durand (2003). Note that in early diagenesis, low molecular weight molecules (LMW) are also formed, in particular biogenic methane in specific environments. At the entry into the metamorphism, where minerals of sediments begin to be deeply transformed, only simple molecules and a carbonaceous residue remain from kerogens. This residue will eventually produce graphite[11]. Source Durand 2003. Courtesy of Editions Technip.

biochemical degradation by microorganisms still present in the sediment. Kerogen then loses much of the nitrogen, which it had inherited from biological molecules. This step sometimes leads to the development of methanogenic bacteria, where the sediment does not contain oxygen, sulphates or nitrates. They produce biogenic methane (bacterial, biochemical) from organic substances dissolved in the porosity of the sediments. The production mechanisms are the same as those that produce marsh gas, cemeteries will-o'-the-wisps, or biogas in agricultural fermenters. This phenomenon may occur in freshly deposited sediments up to depths of up to a few hundred meters.

This stage is followed by progressive pyrolysis under the effect of the increase in temperature, which accompanies burial.

This pyrolysis (Figure 11) first produces water and carbon dioxide, this is the stage called diagenesis. It corresponds to a decrease in the oxygen content of the kerogen. Then comes the so-called catagenesis stage, where oil, mainly hydrocarbons, but also to a lesser extent more complex heteroatomic molecules, resins and asphaltenes (see Chapter 1.1), is formed. Then natural gas is formed, that is to say hydrocarbons,

---

[11] The formation of graphite from the carbonaceous residue is not automatic. In particular, strong shear stresses are required in rocks *(Bonijoly et al., 1982)*.

which would be gaseous under surface conditions (see Chapter 1.2). The gases form from the remaining kerogen, but also from the previously formed liquid hydrocarbons. This distinguishes the oil zone from the wet gas zone, which does not mean that water is formed, but that the gas coexists with liquid hydrocarbons. Catagenesis corresponds to a rapid decrease in the hydrogen content of kerogen: the quantity of hydrocarbons formed per unit mass of initial kerogen depends in a first approximation on its initial content of hydrogen. The productivity of hydrocarbons per unit of initial mass thus decreases from type 1, the richest in hydrogen, to type 3, the less rich in hydrogen.

Catagenesis is followed by the metagenesis stage, where residual kerogen, increasingly low in oxygen and hydrogen, and therefore increasingly rich in carbon, coexist in the source-rock with what is called the pyrobitumen. This is the solid residue from the conversion of liquid hydrocarbons into gas, analogous to the coke produced in the coking units of certain petroleum refineries. A little more gas is formed, but is dry, that is to say, containing no more liquid hydrocarbons.

The gas thus formed in depth by this natural pyrolysis is called thermal gas. It is distinguished from the biogenic (bacterial, biochemical) gas, which is sometimes formed in recent sediments, by the presence of hydrocarbons other than methane, and an isotopic composition of its methane carbon different of that of biogenic methane (*Tissot and Welte, 1984*).

Anglo-Saxon petroleum geochemists often use pictorial terms, sometimes with culinary connotation, to designate these different stages: so kerogen is termed immature during diagenesis, mature during catagenesis, this being called oil window. Kerogen in metagenesis is often said overmature or overcooked. Metagenesis is also called gas window or over mature zone. The evolution of kerogen is called maturation or even cooking.

The sequence of events is the faster the higher the temperature gradient.

Methane, the main constituent of natural gas, is a molecule that is stable at high temperature, it is necessary to reach about 1000 °C, in order to achieve a notable rate of decomposition on the laboratory time scale. Of course, such temperatures are not reached in sedimentary basins, but at great depths, greater than 5 or 6000 meters, i.e. at the end of the metagenesis stage, the immense duration of geological time helping, it nevertheless begins to be slowly cracked to give hydrogen and a carbonaceous residue. The hydrogen thus formed is very reactive and therefore recombines with other molecules present in its environment. And if it is found in fairly large quantities in some small deposits (Table 2), there is no known major deposit where it is a major constituent. Nitrogen is also often formed at this stage from nitrogen still remaining in residual kerogen, and this explains why nitrogen may be abundant in once deep natural gas reservoirs.

These transformations depend therefore on temperature and time. At identical depth and age, they will be all the more advanced in a sedimentary basin than the geothermal gradient is high. For the same geothermal gradient, they will be all the more advanced than the depth and/or the age is important. Therefore, the depths shown in Figure 11 are only indicative, and may vary by several km for the same degree of transformation, depending on age and geothermal gradient.

However, when sediment temperatures exceed 200 °C, which corresponds to average conditions at about 6–7 km depth, the potential for hydrocarbon formation from the kerogens contained therein can be considered to be virtually exhausted. Beyond that, the sediments begin to enter the so-called metamorphic zone to be deeply transformed into metamorphic rocks.

These transformations are described and can be modeled in terms of chemical kinetics. Originally conceived by *Tissot (1969)* at the Institut français du pétrole et des énergies nouvelles (IFPEN), these kinetic models have been perfected and are now able to describe the composition of solid, liquid and gaseous products formed as a function of the thermal history of the source-rock according to the type of kerogen that it contains.

Figure 11 corresponds mostly to what happens in a kerogen-containing sediment where fluids it produces can not escape (closed medium). We compare hereunder this case with that of an open medium, from which all the fluids can escape as they are formed, and that of an intermediate medium between open and closed medium, which is the case of actual situations.

Figure 12 shows an example of the results of modeling of these situations with regard to hydrocarbons only.

Three types of schematic situations are represented:

- **a**: it is assumed here that the hydrocarbons formed during burial are expelled integrally as soon as they are formed in their source-rock. This is an open system. The hydrocarbons which form progressively in the source-rock and are immediately expelled are then very largely hydrocarbons with a carbon number greater than 5, that is to say liquid hydrocarbons. The molecules are formed directly from the kerogen by primary cracking and do not remain sufficiently long at the temperature of the reaction medium to be broken down into smaller molecules by secondary cracking.

- **b**: on the contrary, it is assumed that hydrocarbons can not leave the source-rock where they were formed. This is thus a closed system, as in Figure 11. There is then with increasing depth an increasingly greater destruction in the source-rock of the liquid hydrocarbons to form gaseous hydrocarbons by secondary cracking.

- **c and d**: the most common situation is simulated here, where only part of the hydrocarbons formed can be expelled from their source-rock as they are formed, once sufficient saturation of the porous medium has been reached (see Chapter 3.1.2): c shows the evolution of the composition of the hydrocarbons remaining in the source-rock, while **d** shows that of those that escape from it.

The formation of mobile fluids, water, carbon dioxide, liquid and gaseous hydrocarbons, of course has its counterpart in the chemical composition of residual kerogen. The diagram of Figure 13 shows the evolution according to the initial contents of carbon, hydrogen and oxygen (the 3 main constituent atoms of the kerogens), for

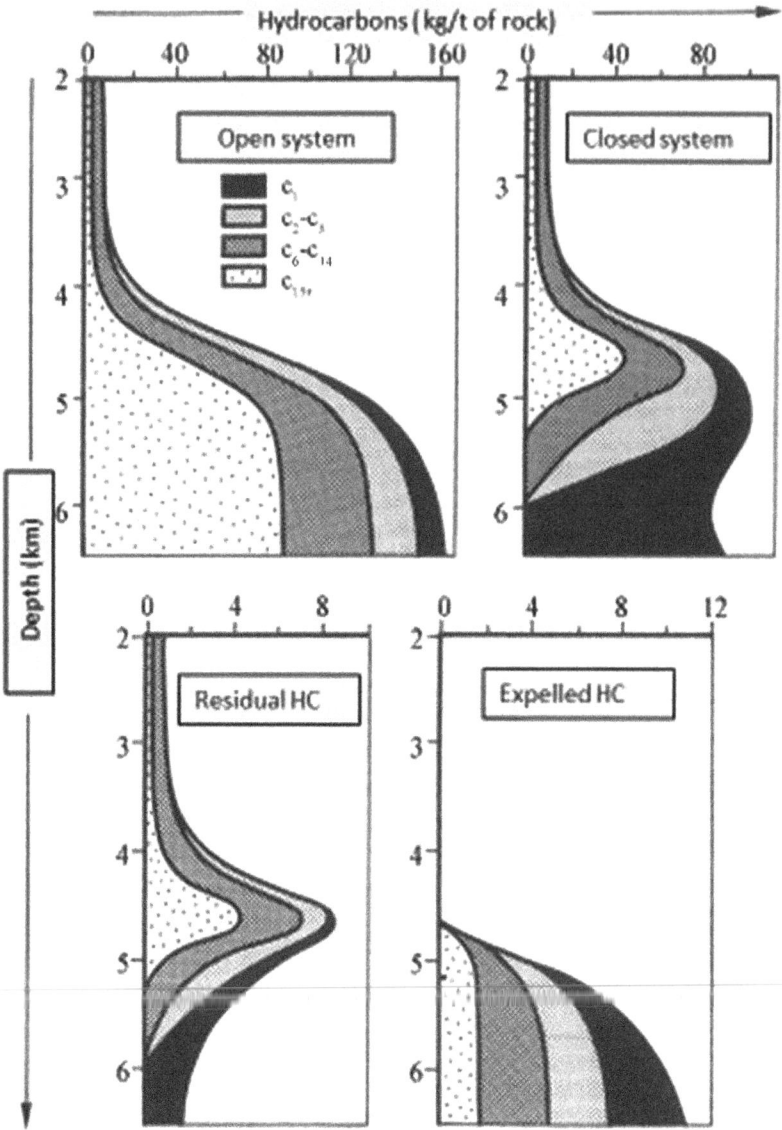

Figure 12 Numerical simulation of the formation of hydrocarbons in a source-rock under different conditions of expulsion of the hydrocarbons formed. The quantities formed and expelled are expressed here in kg of hydrocarbons formed and expelled per ton of rock. The situations a and b correspond to the modeling of a rock very rich in kerogen and with high oil generating potential, and the situations c and d to that of a rock with a petroleum generation potential in the order of 10 times lower than for a and b. According to Ungerer (1993) in Rojey et al. (1997). With the permission of Editions Technip.

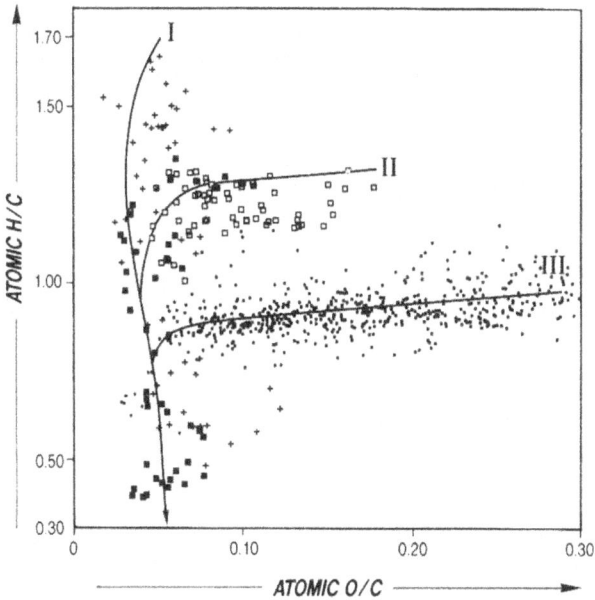

Figure 13 *Maturation pathways during the burial of three reference series of kerogens in a Van Krevelen-Durand diagram: On the abscissa the O/C ratios and on the ordinate the H/C ratios, expressed in atomic ratios, that is to say the ratio of numbers of atoms contained in the kerogen. The black lines converging towards the origin of the coordinates show the mean decreases of these ratios as a function of the "maturity" of the samples acquired during their thermal history. The dispersion of the samples around the mean lines corresponds to variations, within the same type, of the initial sedimentation conditions and the composition of the original biomass. Type I is represented here by samples of the Green River Shales in the United States, Type II by samples of the "schistes-cartons" of the Lower Toarcian of France and Germany, and type III mostly by samples of clays and coals of the coal series of the Mahakam Delta in Indonesia. For each of the three types, diagenesis is characterized by a faster decrease in the O / C ratio than the H / C ratio. It is the reverse for the catagenesis which succeeds it. During the metagenesis genesis, the displacements become very small because the kerogens then lost most of the oxygen and the hydrogen that they originally contained. After Vandenbroucke et al. (1993), Courtesy of Editions Technip.*

three series of samples which belong to the three reference types of kerogens, as described in Chapter 2.1.

For each of the three series, samples were taken in oil exploration wells at various depths, but the origin and composition of the kerogen of these samples were initially substantially the same at the time of deposition of the kerogen containing sediment. Thus, the so-called maturation (evolution) pathways of the different types of kerogen, that is to say, in each series, here the changes in C, H and O composition of kerogen, are observed as a function of burial, and more precisely the maturation stage of each sample, which is determined by its thermal history.

This diagram also gives an initial indication of the petroleum and gas generating potential of kerogens, i.e. the total amount of oil and gas that they can produce per unit of initial mass during burial. This potential is in fact mainly linked to the abundance per unit mass of CH chains known as alkyl or aliphatic chains, which can be identified in particular by infrared spectroscopy (*Rouxhet et al., 1980*), which will be the most hydrocarbon-generating during thermal cracking. This abundance decreases from type 1 to type 3. The ratio H/C, which is all the higher as these structures are abundant, is an approximation.

3

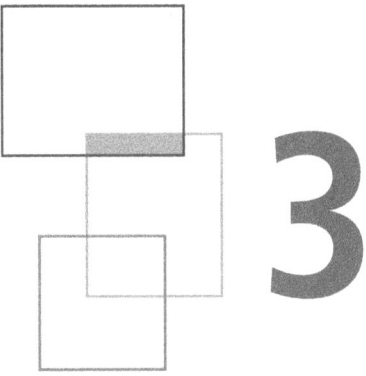

How are their deposits formed?

3.1 Fluid fuels: oils and gases

3.1.1 Pressure: a major role in the formation of oil and gas fields

Increasing burial results in increased mechanical stresses on sediment minerals, on kerogen, if any, and on the fluids (water, hydrocarbons, etc.) contained in their pores since the sediments have practically always an important porosity (exceptions are evaporites: salt, potash, sulphates...). They have, however, very variable permeabilities to fluids.

Contrary to widespread opinion, pressure has little effect on the formation of fossil fuels. But pressure gradients have a lot to do with the movement of fluids in sedimentary basins and therefore in the formation of oil and gas deposits. It also has on the compactness and anisotropy of sediments in general, and coals in particular.

The so-called geostatic vertical stress S exerted on a sediment is due to the weight of the overlying sedimentary column. It therefore depends on the height of this column and the average sediment density in the column, i.e. about 15 to 25 MPa[12] per km depending on the situation. It is distributed, as a first approximation, between the minerals and the fluids of the sediments according to a law called the Terzaghi's law

---

[12] Pascal (Pa) is the unit of pressure of the International System of Units (SI). It corresponds to a force of 1 newton exerting on a surface of 1 m^2. 1 megapascal (MPa) is in old units 10 bars and therefore a little less than 10 atmospheres.

or the Effective Stress law: $S = \sigma + P$, where σ is the stress actually exerted on the minerals and P is the pressure of the fluids contained in the porosity of the sediment.

The sedimentary rocks belong to four broad categories according to their mineralogy:

- Sandstones and silts, formed from sand with predominantly quartz and feldspars, with very variable grain sizes, the silts having very fine grains.
- Limestones and dolomites, composed of carbonates.
- Argillites and marls, the first composed mainly of clay minerals, the latter also containing carbonates.
- Evaporites (salt, potash, sulphates, etc.).

The evaporites are formed by evaporation of the sea water, which causes them to precipitate from ions present in the latter. They are not porous and therefore contain very little water, compact very little and are particularly waterproof. Chlorides (salt itself = sodium chloride, potash = potassium chloride, rarer), are plastic, of lower density than other types of rock and fluent under geostatic stress. They can cross the overlying sediments, forming intrusions of globular form sometimes gigantic, called diapirs. The sulphates (gypsum) are transformed in depth by dehydration in anhydrite, massive rock very hard, dense and impermeable. Never containing kerogen, evaporites cannot be source- rocks. Their impermeability makes them excellent covers for oil and gas fields (see below).

The sandstones are very porous and permeable. Their minerals are not very deformable and are therefore resistant to the stress exerted by the overlying sedimentary column: the effective stress on the minerals is therefore maximum and the fluids contained in their pores are therefore subjected essentially to the pressure of the column of water located above them, i.e. a pressure mostly close to the hydrostatic pressure. These rocks compact little with stress.

Carbonates have porosities and permeabilities very variable according to their origin, and often a porosity of cracks. As for sandstones, their minerals are not very deformable and therefore bear a maximum effective stress.

The clays (clay minerals-rich rocks) and the marls (mixtures of clays and carbonates) are formed of minerals of extremely small size. They have large porosities, although their pores are very small, and contain a lot of water, but their permeabilities are very low. Moreover, their minerals are deformable under stress. Their fluids thus support a greater part of the stress exerted by the sediments than in the preceding cases, and are therefore at a pressure higher than the hydrostatic pressure. They are thus progressively expelled in cracks and faults, or permeable rocks with less fluid pressure (sandstones, porous or cracked limestones), and consequently a clay or a marl compacts a lot, i.e. decreases in volume, during its burial. Recall that these rocks are also those that can contain kerogen and therefore are likely to become oil and gas source-rocks.

To better understand this phenomenon, let us consider a very simple model in the form of a cylinder full of water, the bottom of which is filled with undeformable

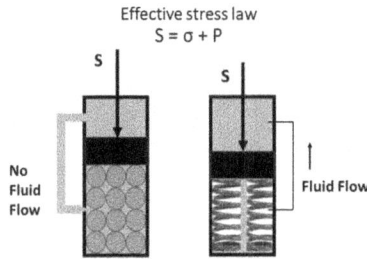

Effective stress law
S = σ + P

S

S

No
Fluid
Flow

Fluid Flow

Figure 14   *Schematic models of sandstone, left, and argililite, on the right.*

glass or metal balls (Figure 14). On the balls is supported a solid piston which is perfectly sealed and movable without friction, the weight of which is supposed to represent the stress of the overlying sediments. On the side of the test piece, at the level of the balls, there is a large diameter bypass side tube joining the bottom of the test piece to its part situated above the piston: the resistance of the balls opposes the descent of the piston and the pore water does not support the weight. It only supports the weight of the water column located above it via the bypass tube and is therefore at hydrostatic pressure: there is no water flow. The effective stress exerted on the balls is therefore equal to the pressure exerted on them by the piston minus the hydrostatic pressure. It is the model of a sandstone. Now replace the balls with springs that are not firm enough to oppose piston pressure. These springs are then compressed, and the interstitial water in this case supports a larger part of the weight of the piston, while the stress exerted on the springs is lower than that exerted on the rigid balls. Its pressure is then higher than the hydrostatic pressure, it escapes upwards through the bypass tube. The piston descends until the springs, by compacting, end up bearing its weight. It is the model of an argillite.

To complement these models, one must also take into account the permeability to water, which is the capacity of the medium to the circulation of it. This permeability is much more important in the case of sandstones than in the case of clays, which are, as everyone knows, very impermeable rocks. These differences will be simulated by a branch tube of large diameter and therefore without substantially loss of charge in the case of the balls, and of very small diameter, a capillary of very great pressure loss, in the case of springs.

While the large-diameter tube hardly opposes resistance to the circulation of water, the same is not true of the capillary, where the rate of flow of the water will depend on this resistance, that is to say the permeability, and the pressure gradient in the capillary. This is expressed by the Darcy's law: $dQ/dt = -k.\Delta P$, dQ/dt being the quantity of water expelled per unit of time, that is to say its flow rate, k a coefficient representing the permeability of the capillary (of the rock) to water, and ΔP the pressure difference between the ends of the capillary.

This permeability regulates the velocity of water flow, and thus also the rate of compaction of the "clay".

Heat and pressure also cause changes in sediment minerals. This is called mineral diagenesis. The effects can be important: in particular dissolutions or cementations can occur, which modify the porous network of rocks, specially sandstones and carbonates. However, these effects are much less important than those of metamorphism, which profoundly transforms minerals from rocks, but which corresponds to depths and hence higher temperatures and pressures than those, which preside over the formation of fossil fuels and their deposits.

3.1.2 *From source rocks to fields: expulsion from source-rocks (primary migration) and accumulation in reservoirs (secondary migration)*

The sediments are, we recall, rocks deposited in an aqueous medium. Water, in most cases sea water, always saturates from the outset their porosity when it exists.

However, hydrocarbons are insoluble in water, with the exception of a few, mostly methane (CH_4), ethane (C_2H_6), and benzene (C_6H_6). Methane is soluble up to about 20 grams per cubic meter of water while the two others have even lower solubility. Resins and asphaltenes are insoluble. The displacement of oil and gas in the sedimentary basins can therefore only take place in phases, which are distinct of the water phase.

This observation leads us to take into account forces which we have not yet discussed, the capillary forces: indeed, there exists naturally between the water and hydrocarbon phases when they are present in a porous medium a capillary tension, a force which is linked to the repulsion between the molecules of the water and the molecules of hydrocarbons. Figure 15 shows what happens in a porous medium occupied by two fluids, which are insoluble in each other and finely divided. One of the two has

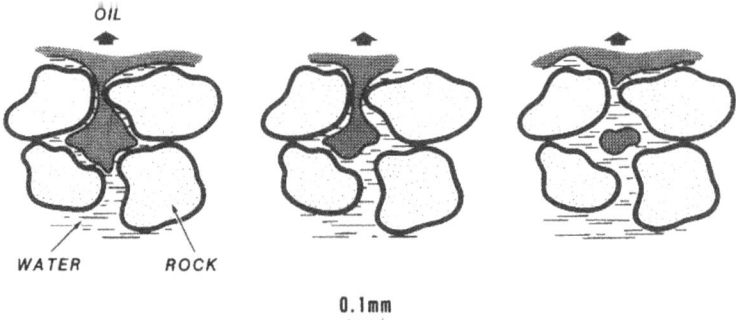

OIL

WATER   ROCK

0.1mm

Figure 15 *Capillarity and capillary pressure in porous media. Here is shown a sketch of three successive steps of the moving of an oil stringer in a porous medium filled by water. Source Durand (1988). Courtesy of Editions Technip.*

more affinity than the other for mineral surfaces, it is said to be wettability. This is generally the case of water, hydrocarbons being wetting only for certain carbonates. They are also for kerogen. In order to displace drops of hydrocarbons, which tend to be in the center of the pores, and for them to pass from one pore to the other, these droplets have to be deformed to overcome the so-called capillary pressure, the value of which is proportional to the capillary tension and inversely proportional to the size of the pore thresholds of the sediments. This is Laplace's Law: $P = 2\gamma/\ r$, P being the capillary pressure, γ the capillary tension and r the radius of the pore threshold.

To cause hydrocarbon phases to flow through the pores of sediments which are initially filled with water, their pressure must therefore be greater than that of the water by a value at least equal to the capillary pressure.

The saturation of the fluid phases, that is to say the proportion which they occupy of the porous space of the rock which they permeate, is a very important parameter: the flow of a phase is all the faster as its saturation is important.

It is then understood that it becomes increasingly difficult for hydrocarbons to flow when their saturation in the porous medium decreases. This is what happens when a hydrocarbon field is exploited, and that is one reason for the low oil recovery rate of most oil fields.

The source-rocks, once again, are mainly clays or marls: these are porous rocks, but because of the extreme smallness of their minerals the dimensions of their pore constrictions are extremely small and they are very poorly permeable to fluids. Laplace's law implies that the hydrocarbons once formed should not be able to leave the source-rock, due to a much too high capillary pressure.

Yet, as shown in the sketch in Figure 16, it becomes possible once a sufficient amount of oil has formed in the source-rock: the latter is gradually buried, but for a long time there are no hydrocarbons formed within it. The water it contains is expelled slowly, as described above by Darcy's Law. There is little left over when kerogen begins to form hydrocarbons by thermal cracking. These first constitute a discontinuous phase in contact with the laminae and the clumps of kerogen, but the capillary pressure due to the presence of residual water opposes their displacement. Hydrocarbons cannot be expelled, while water continues to be expelled.

The amount of hydrocarbons gradually increases during the subsequent burial, while their saturation in the porous medium, that is to say the ratio of their quantity to the quantity of residual water, increases. At the same time, they support an increasing amount of geostatic stress as fluids, and their pressure increases.

They then end up creating in certain parts of the bedrock continuous phases under high pressure, which no longer encounter the obstacle of capillary pressure to oppose their movement. They are then expelled and injected into much more permeable drains, cracks, faults or porous rocks (sandstones, carbonates) where the fluid pressure is hydrostatic or in any case lower than in the source-rock. The expulsion becomes possible once sufficient saturation of the porous medium is reached and becomes more and more rapid, as explained in Figure 16.

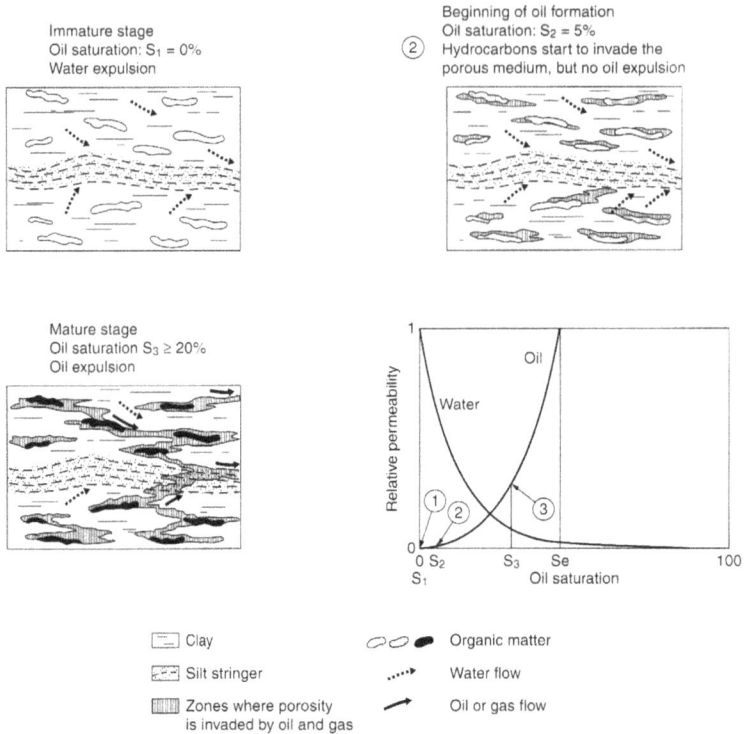

Immature stage
Oil saturation: $S_1 = 0\%$
Water expulsion

Beginning of oil formation
Oil saturation: $S_2 = 5\%$
② Hydrocarbons start to invade the
porous medium, but no oil expulsion

Mature stage
Oil saturation $S_3 \geq 20\%$
Oil expulsion

Relative permeability

Oil

Water

① ② ③

0 S_2 S_3 S_e 100
S_1 Oil saturation

Clay

Silt stringer

Zones where porosity
is invaded by oil and gas

Organic matter

Water flow

Oil or gas flow

Figure 16 *Sketch showing the mechanisms of hydrocarbon expulsion from a source-rock: 1: during diagenesis, there is no production of oil and gas in the source-rock. The water contained in the porosity is expelled during compaction; 2: At the beginning of catagenesis, there is not enough oil formed to constitute a continuous phase in the source-rock. The capillary pressure prevents its expulsion, but the water continues to be expelled; 3: Once a continuous phase is formed in the source rock, the oil begins to be expelled. Sufficient average saturation of the pores of the bedrock, S_3 on the Figure, must be achieved, as reflected by the oil and water permeability curves in the inset. This saturation may increase up to S_e, i.e. a nearly monophasic expulsion of the oil. This implies that expulsion occurs earlier in kerogen rich rocks than in poor rocks, and therefore that the expelled products are richer in gas for the latter than for the former. Source: Ungerer, in Bordenave et al. (1993). Courtesy of Editions Technip.*

This expulsion of hydrocarbons from the source-rock is called primary migration.

Macroscopically, the expulsion of water and hydrocarbons over time can be described by the so-called polyphasic flow laws in porous media, which are empirical extensions of Darcy's Law cited above. As in Darcy's law established for a single fluid phase (monophasic), the velocity of the flow is governed by the pressure gradient and the permeability of the medium to the fluid phase, but a relative permeability is assigned to each phase, which depends inter alia on its saturation, and its viscosity, in the porous medium. It is found that the higher the saturation of one of the phases, the higher its relative permeability, and hence its flow velocity (Figure 16 in the lower right) for the same pressure gradient. Between the phases, there is a pressure difference equal to the capillary pressure.

However, this expulsion is not complete, as simulated in Figure 12c. As will be discussed further on, non-expelled quantities remaining in the source-rock can sometimes be exploited in the form of so-called shale oil and shale gas.

The density of petroleum or gas phases with respect to water is exceptionally (some bitumens and extra-heavy oils) higher than that of water. The latter is omnipresent in most of the rocks of the sedimentary basins. The oil and gas phases are therefore subjected to an upwardly directed Archimedes' buoyancy: this is proportional to the difference in density between water and oil or gas phases, and to the height of the hydrocarbon phases that these constitute. The density difference with water is of the order of 0.1 to 0.3 in the case of the petroleum phases and of 0.6 to 0.9 in the case of the gas phases. This buoyancy is negligible in front of the capillary pressure in the very fine grain rocks and hence with very fine pores which are the source-rocks and therefore can not contribute to expulsion from them, at least as long as the phases are discontinuous. However, it becomes more important than this capillary pressure in drains with high permeability, cracks, faults and coarse-grained rocks such as porous sandstones or limestones, as soon as the height of the oil and gas phases originating from the source-rocks reaches a few meters. There follows an upward movement of oil and gas, thanks to these drains. This is called secondary migration.

The normal destiny of the oils and gases expelled from the source-rocks is therefore to dissipate very slowly towards the surface of the sedimentary basins where they were formed. These "leaks" can be sometimes large enough to constitute on the surface what are called oil or gas shows, witnesses of the formation of oil and gas in the depths of a basin.

But it may happen that a part of them is on their ascending path stored progressively in porous and permeable rocks, sandstones or carbonates, called reservoir rocks or reservoirs, where they have been trapped in a geological structure closed by a barrier of permeability. It is often made up of a clay rock, sometimes salt, or what is called a fault closure. An inverted image is that of impermeable bottom depressions that trap runoff water to form ponds and lakes.

Thus, the exploitable accumulations of oil and natural gas are formed, which are in fact provisional local accumulations, on a geological time scale.

The traps are of a very varied nature, size and geometry: Figure 17 shows the most common.

The quantity of oil and gas thus trapped temporarily in structured, exploitable reservoirs represents only a very small fraction of the quantities formed in the sedimentary basins, probably only around 1% globally. A part of the quantities formed also remains, as we have seen, provisionally trapped in the source-rocks: a fraction of which the importance is not well understood for the moment, known as shale oil

a

Impermeable rock

Gas

Oil

Water

Permeable reservoir
(Sandstone)

b

Impermeable rock

Gas

Water

Fault

Permeable reservoir
(Sandstone)

c

Salt

Oil Gas Permeable
Impermeable reservoir
rock (Sandstone)

Water

d

Impermeable rock Gas

Permeable reservoir
(Sandstone)

e

Impermeable rock

Gas

Water Oil

Permeable reservoir
(Fissured limestone)

f

Impermeable rock

Oil Water

Permeable reservoir
(Fissured limestone)

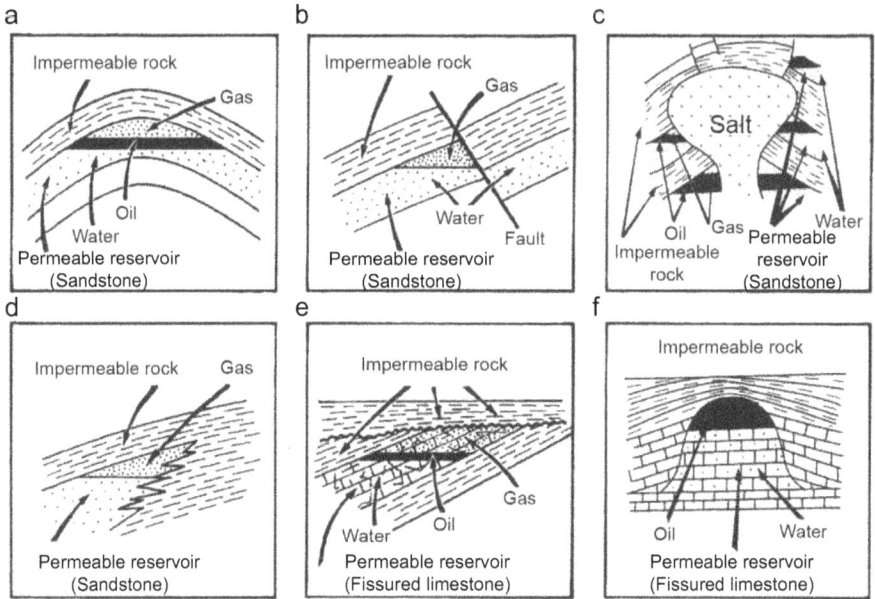

Figure 17 *Some types of traps: a: anticline; b: fault closure; c: trapping along a salt dome (note: salt, which is a plastic rock less dense than the rocks that overcome it, deforms and rises very slowly towards the surface by piercing these rocks, thus creating traps); d: stratigraphic bevel; e: trap under unconformity; f: limestone reef. No scale is indicated, as these traps vary widely in size.*

and shale gas, is recoverable. But most of the oil and gas formed in the sedimentary basins dissipates slowly during geological time without being trapped in exploitable accumulations.

Here we see the very low yield of the mechanisms that lead from photosynthesis by plants to the formation of kerogens and then to the exploitable accumulations of oil and gas (Figure 18): the energy efficiency of photosynthesis is of the order of 1% of the solar energy received, the yield of the formation of kerogens from organic debris is about 0.1%, the average yield of the transformation of kerogens into petroleum and gas is in the order of 10%, in view of the fact that a large part of kerogens will not undergo sufficient temperatures during its geological history to produce significantly oil and gas, and the quantity of oil and gas which can be exploited is in the order of 1% of the quantities produced in sedimentary basins. The overall efficiency of the chain, which goes from solar energy to the exploitable accumulations of oil and gas, is therefore only around 10^{-8}! According to *Huc (1980)*, the quantities of organic carbon stored annually in biomass would be in the order of 100 billion tons, about half of which would come from plankton, the main source of the most prolific kerogens in oil and gas, and half terrestrial plants, the principal providers of coals. In the order of 500 tons of oil and gas, on average, each year would be added to the existing reserves

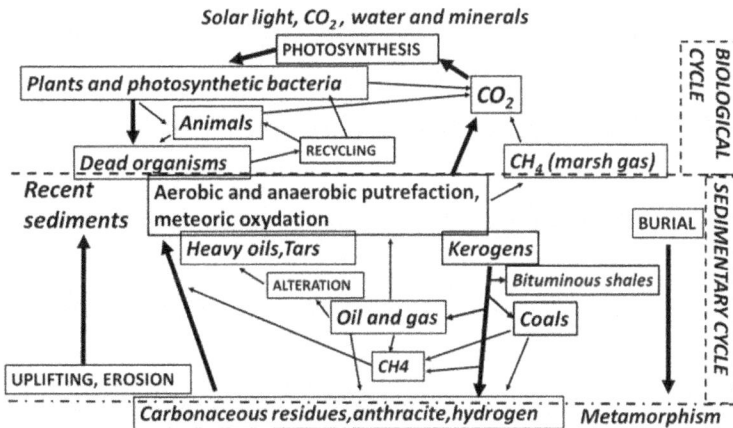

Figure 18 *A representation of the biological and sedimentary cycles of carbon in its organic form.*

of oil and gas, whereas in 2015 we have consumed nearly 8 billion tons of those. The situation is better for coal, which remains in place and for which there is therefore no loss per migration. But only coals shallower than about 1500 meters are accessible by mines. It can be estimated that the conversion yield to reserves here is about 10 times greater than the previous one, i.e. 10^{-7}, for a quantity of kerogen sedimented each year which is about the same. This time around 5000 tons of reserves would be created every year, for consumption presently of around 8 billion tons. These calculations are very approximate, but they highlight the enormous gap between the natural production rate of fossil fuel reserves and the rate at which they are consumed by man, which is of the order of 1 million to 10 million times higher!

A part of the oil and gas formed, which has not been able to leave the source-rocks is sometimes exploitable: this is so-called shale oil and shale gas. But this only insignificantly improves the performance indicated above, at least for the time being.

The petroleum and gas phases, of lower density than the water phase, lie above them in the reservoir and exert on its roof an Archimedes thrust. Figure 19 diagrammatically shows the different modes of association of the oil and the gas as a function of their composition and the temperature and pressure conditions in a conventional reservoir of anticline type.

3.1.3 Transformations in fields

Once accumulated in their deposits, the geological history of oil and gas is not over. They may in fact leave their reservoir, either laterally at the base of the reservoir when it is full, or by slowly escaping from above due to a defect in the cover, cracks or insufficient impermeability. This is called tertiary migration or dysmigration.

(1) Oil without gas

Cover
Oil
Aquifer

(4) Gas saturated with oil and oil ring

Gas
Oil
Gas
Oil

(2) Oil sub-saturated with oil

Oil
Gas

(5) gas sub-saturated with oil

Gas
Oil

(3) Oil saturated with gas and gas-cap

Gas
Oil
Gas

(6) Dry gas (without oil)

Gas

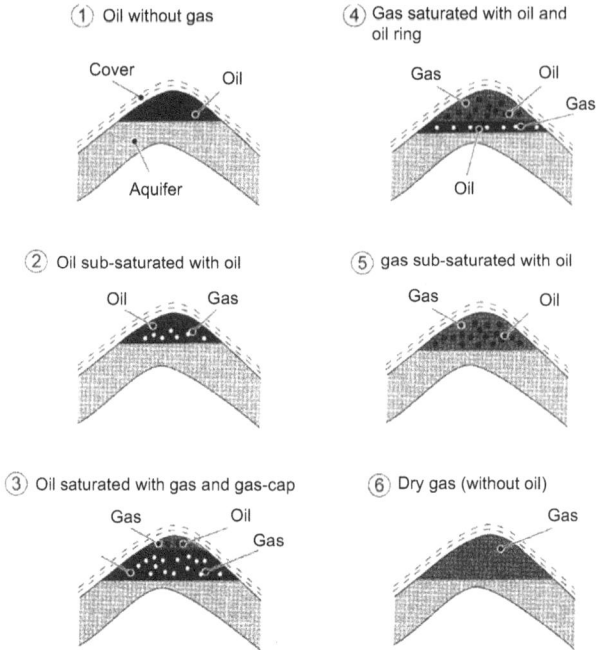

Figure 19 *The different modes of association[13] for oil and gas in their fields: 1: oil without gas; 2: oil sub-saturated with gas; 3: oil saturated with gas and gas-cap; 4: gas saturated with oil and oil ring; 5: gas sub-saturated with oil; 6: dry gas (without oil). The gas dissolved in the oil is called associated gas; The oil dissolved in the gas and which is recovered at the wellhead is called condensate. This Figure is similar to Figure 4.*

Dysmigration can lead to the formation of new deposits, but also to surface leakage, and thus to the formation of oil or gas shows.

Inside the reservoir the oil can be converted by thermal cracking into lighter oil and pyrobitumen, provided that the reservoir temperature and/or time are sufficient. This is called dismutation. Oil and gas can also react with the minerals of the surrounding rocks to form carbon dioxide, or sulfur compounds, particularly hydrogen sulphide. This phenomenon exists in deep reservoirs, therefore at high temperature, when they are made up of carbonates with the presence of anhydrite (anhydrous calcium sulphate) (*Goldstein and Aisenshtat, 1994*). The gas from the Lacq deposit, the largest

---

[13] In addition to the situations shown in Figure 19, there is an intermediate situation between gas under-saturated with oil and dry gas, which is called wet gas. Unlike dry gas, it contains some little liquid hydrocarbons recoverable in gas processing plants. Biogenic gas (biochemical, bacterial) does not contain any other hydrocarbon than methane and is never associated with oil in its deposits. It is therefore currently dry gas. However, its deposits may sometimes contain thermal gas from a deeper source.

in France, is in this case and originally contained 9% of carbon dioxide and 15% of hydrogen sulphide (Table 2).

The introduction of gas or light hydrocarbons into an oil reservoir can cause deasphalting, i.e. precipitation of asphaltenes-rich solid bitumen, and the formation of a lighter oil in a similar manner to what happens when asphaltenes are separated from oil in the laboratory (see Chapter 1).

Reservoirs may also be uplifted near the surface due to tectonic inversion and accompanying surface sediment erosion. Their contents are then altered by evaporation, leaching (by meteoric fresh waters or by aquifers laden with salts, brines), oxidation and bacterial alteration, and gives rise to accumulations of extra-heavy oils or bitumens (Figure 20).

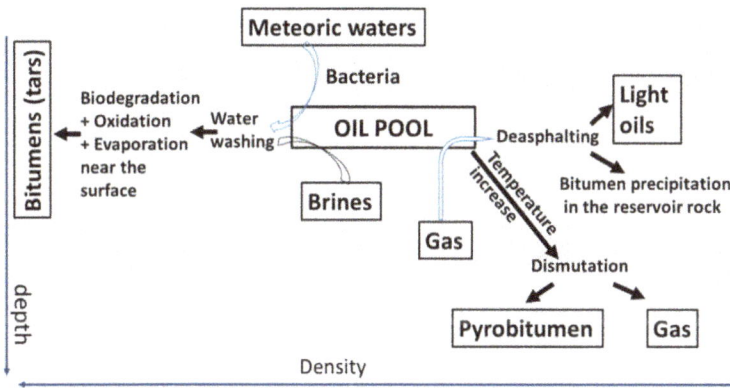

Figure 20 *Transformations of oils in their reservoirs. As indicated by the arrow of densities, deasphalting and dismutation are phenomena giving rise to the formation of hydrocarbon phases of lower density than the initial oils. As indicated by the depth arrow, dismutation takes place at depth. Leaching, evaporation, bacterial action and oxidation occur at shallow depth and cause an increase in density of the hydrocarbon phases. After Connan (1984).*

This is the way that the gigantic deposits of extra-heavy oils of the Orinoco tar belt in Venezuela and of tar (bitumen) sands of the Athabasca in Canada were formed.

Each deposit has a complex history of its own. There is a wide variety of situations, and detailed geological and geochemical studies are required to untangle them.

The exploitation of these deposits, that is to say the recovery of the oil and the gas which they contain, must deal with the same physical laws as those presiding over their constitution, that is to say those of the polyphasic flow of the fluids in a porous medium (Figure 16). The flow of each phase, water, oil, gas can be described by a flow law, where the flow depends on the pressure differences and the relative permeabilities of the porous medium at these phases. It is also necessary to take into account the capillary pressure between phases and the wettability of the mineral surfaces by these phases and therefore the mineralogical and petrophysical characteristics of the deposit.

It should be noted that when deposits such as those shown in Figure 17 are pierced by boreholes, initially the oil and/or gas will rise towards the surface under the effect of the Archimedes buoyancy which is due to their difference in density with the water phase they overcome, but also their decompression, in particular for the gas phase. In a way, the water progressively reoccupies the space from which it was driven away by oil and gas, and tends to rise towards the surface up to the hydrostatic level. This is called natural drainage or primary recovery. Secondly, however, it will be necessary to help the oil and gas to rise towards the surface by injecting water into the aquifer to maintain its pressure. This is called secondary recovery. Then are used more sophisticated techniques, called tertiary or enhanced recovery, to recover the maximum possible of oil or gas. For instance, in order to reduce the viscosity of the oil phase, hence increasing its flow speed, oil can be heated by injection of hot water steam, or mixed with a so-called miscible gas. However, this is costful and practiced on a limited number of fields so far (see for instance *Höök et al., 2013*).

It should be noted that, despite all these efforts, the recovery rate of the oil phase, i.e. the ratio of the quantity finally recovered to the quantity initially in place, if it varies greatly from one field to another, is on the order of only 30%. This is due to the physics of the flows in a porous medium, the recovery of an oil phase depending very much on its saturation, that is to say its proportion in the fluids contained in the pores of the reservoir-rock, on its viscosity and on its wetting of minerals surfaces. Then, the saturation of oil declines as it is extracted, and its viscosity is often important. The average recovery rate for gas, much less viscous, is much better, on average 80%.

There are many deposits, for example most of those of extra-heavy oils and bitumens, or those of shale (source-rock) oils and gases, which as we shall see can be exploited only with unconventional recovery techniques, as opposed to conventional ones mentioned above.

All this is in very complex practice, and each deposit, or almost, is a special case.

3.1.4 *Two examples of the formation of fields*

3.1.4.1 The Viking Graben in the North Sea

The sketch in Figure 21 is inspired by the case of the rift valley in the center of the North Sea, the Viking Graben: it shows the history of the progressive formation of oil and gas deposits in a sedimentary basin.

Sand is first deposited, which gradually transforms into sandstone, and then above is a clay rock (argillite), the base of which is rich in kerogen. The basin deepens and new strata of sediments are deposited and enter successively diagenesis (D on the Figure), catagenesis (C) and metagenesis (M) zones. Oil then gas form

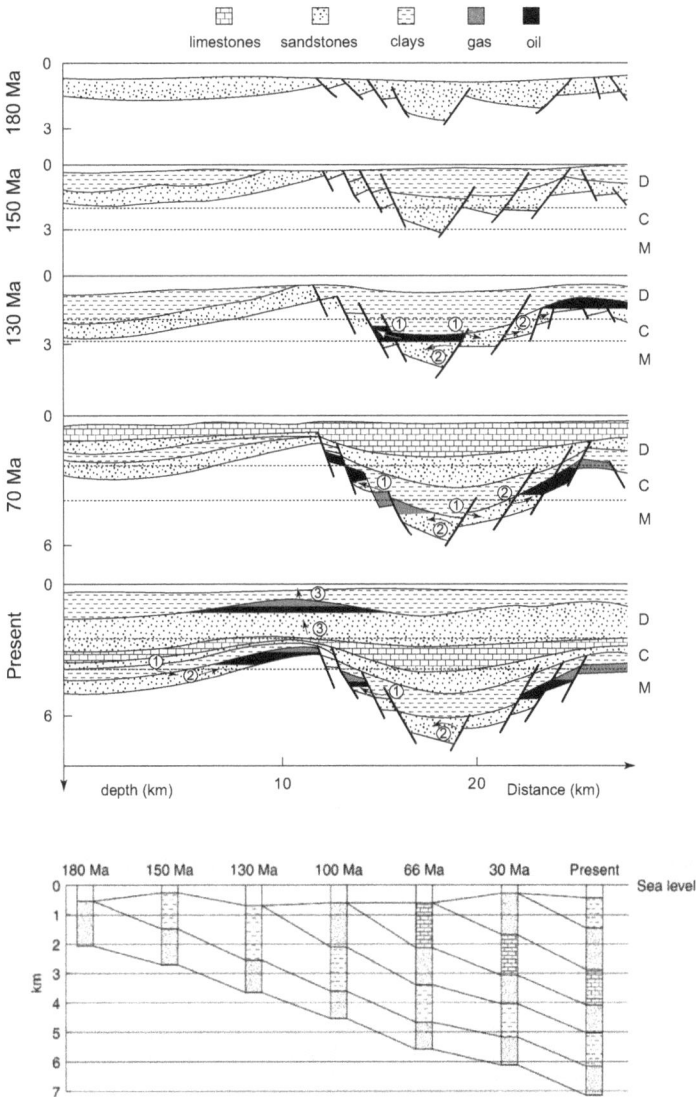

Figure 21   *Sketch showing the gradual formation of oil and gas fields in a sedimentary basin, inspired from the case of the Viking Graben in the North Sea. The upper part of the Figure shows successive stages of the evolution of the basin and that of an argillite rich in kerogen, shown in dashed lines, buried at increasing depths and therefore at increasing temperatures. The oil and gas formed in this source-rock in steps C (catagenesis) and M (metagenesis) were expelled (arrows 1) to drains which led them (arrows 2) to accumulate in traps and to form pools. A part escaped from these traps (arrows 3) to go higher up to form other deposits. The lower part of the Figure shows the evolution over time of the stacking and burial of the sedimentary strata in a fictitious well located in the center of the basin. Source: Durand (2003).*

progressively during catagenesis (C) at the base of the argillite from the kerogen it contains. This argillite is now a source-rock. Then these oil and gas are expelled from it (primary migration), here downwards and laterally in the underlying sandstone because they have no outlet from the top. This expulsion results from their excess pressure with respect to the hydrostatic pressure as explained in Chapter 3.1.1. They then move in an upward movement under the effect of the Archimedes' buoyancy to traps (secondary migration). A part escapes from one of these to form a secondary deposit and then a surface show (tertiary migration). The history of the formation of deposits extends over 180 million years. The depth reached by the source-rock in the center of the basin is a little over 6 km, which corresponds to a temperature of about 200 °C.

3.1.4.2 The Neuquen Basin in Argentina

Figure 22 is an east-west section of the Neuquen Basin in southern Argentina. The complexity of the sediment strata arrangement, as in many other sedimentary basins, is observed. On a basement of Palaeozoic age were deposited sediments of Triassic age, then of Jurassic and Cretaceous ages. These first deposits were then compressed and underwent a tectonic inversion: their upper part was then eroded, as indicated by a gap in the succession of sediments and a discordance (dotted on the Figure). Then a Cenozoic (Tertiary) sedimentary basin settled above this erosion surface. The set then underwent a new tectonic inversion and the sediments thus brought to the surface are now being eroded.

A remarkable fact is the recent (geologically speaking) overloading by a thrust belt, and folding and fracturation of sediment strata in the western part of the sedimentary complex, due to the formation of the Andes. There are also numerous volcanic intrusions, some of which are very recent, as they have passed through the entire stack of sediments.

The formation Vaca Muerta is a formation of a few hundred meters deposited in twenty million years from the late Jurassic to the early Cretaceous. It includes kerogen-rich marls marked "Black shale" in Figure 22, whose current depth from the ground varies from 2000 to 3000 meters.

Figure 23 is a simplified mapping of these Black shales' "maturity" showing the geographical extension of the oil window (catagenesis) and of the dry gas window (metagenesis). There is a growing maturity from East to West. This is probably due to the overloading of the western part of the bassin by the Andean thrust, creating a high thermal flux and a sedimentary thickening which has now disappeared due to erosion.

These Black shales have therefore produced oil and gas. This source-rock has also retained in it a good part of the oil and gas it produced. It is one of the main objectives for the production of shale oil and shale gas in the world outside the United States.

These two examples provide only a very small glimpse of the vast variety of stories of the formation of oil and gas fields in sedimentary basins, and the complexity that they may present. The cooperation of many professionals is necessary to understand

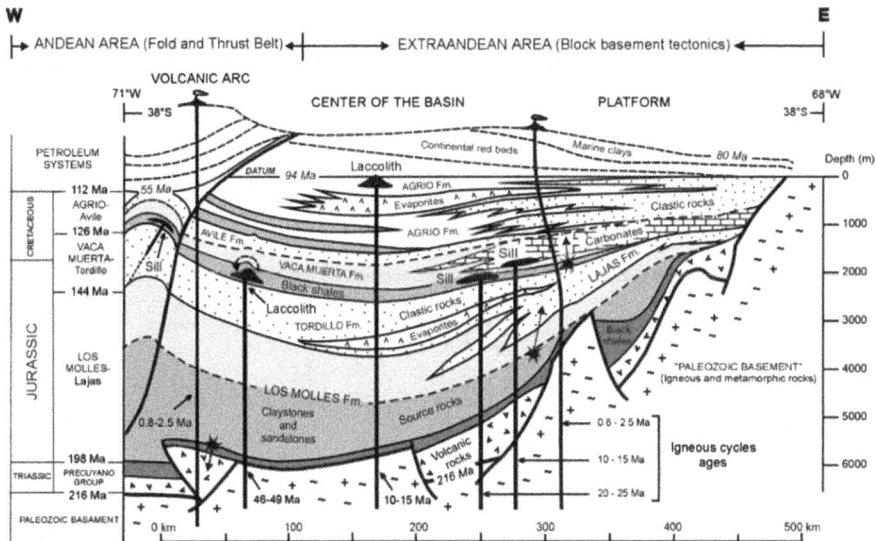

Figure 22 *East-West section of the Neuquen Basin showing the location of the Black shales of the Vaca Muerta formation. Source Wikipedia.*

them. This work, more and more indispensable to the discovery of new deposits, is entrusted to "explorers", geologists, geophysicists, geochemists.

3.1.5   *Biogenic gas, gas hydrates, and dissolved gas in deep aquifers*

Large quantities of biogenic gas can be formed in recent sediments, provided that they provide a sufficiently porous space for the development of methanogenic bacteria, that the sediment water contains dissolved organic substances, and that it is devoid of oxygen, sulphates and nitrates *(Rice, 1992)*. However, they can only form significant accumulation if their zone of formation is already surmounted by an impermeable cover. These conditions have been met, for example, in the river Po plain in Italy, where there are large fields of biogenic gas in turbidites formed of sand lenses packed in clays. Turbidites are sediments formed by subsea landslides of sediment accumulations resulting from the erosion of nearby mountains, here the Alps. The gas accumulates in the upper parts of the sandy lenses and cannot escape because of the clays that pack them. The deposits thus formed were then buried at considerable depths. The Antonella deposit, for example, is currently about 4000 meters deep. In some Western Siberia deposits in Russia, such as Urengoy (Table 2), the biogenic gas contains some thermal gas coming from the deep.

Biogenic gas formation might have been responsible for 20% of the present world reserves of conventional gas deposit *(Clayton, 1992; Rice, 1992)*.

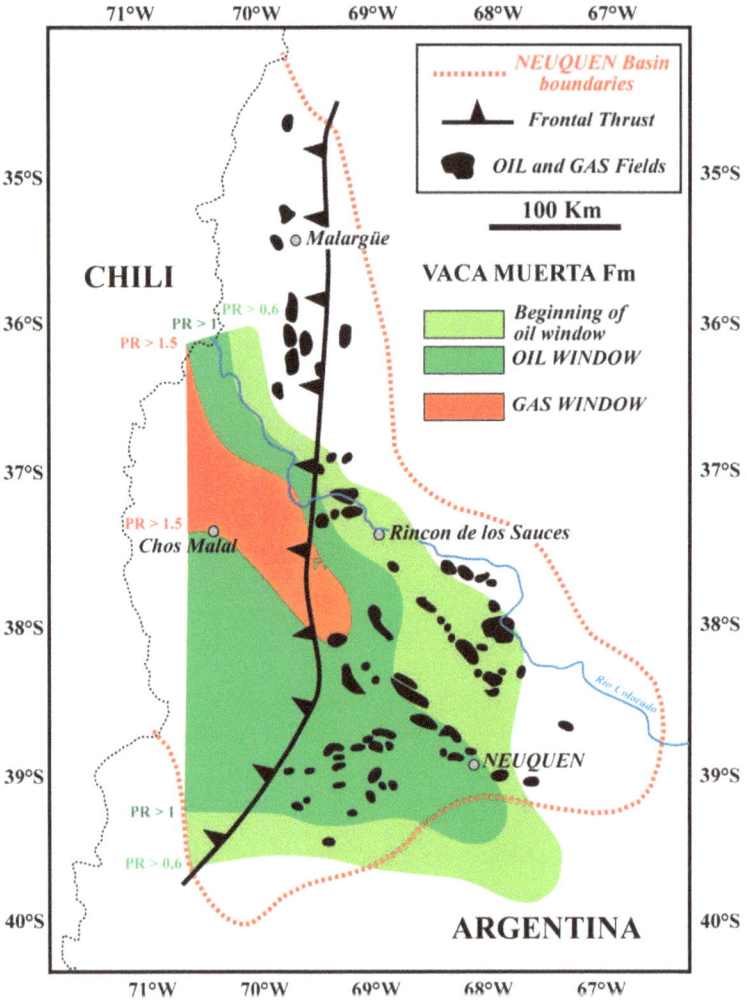

Figure 23 *Geography of the Black shales maturity of the formation called Vaca Muerta in the Neuquen Basin in Argentina: this map shows the extension in this formation of the oil window (in light green and dark green), and of the dry gas window (red). The Vaca Muerta has retained a good part of the hydrocarbons it produced: it is the main objective of possible shale oil and shale gas operations in Argentina. The zones of increasing maturity are marked here with an indicator, the reflectance of vitrinite, denoted PR, whose value increases with maturity. This standardized method is described, for example by Stach et al. (1982). Note that the oil pools indicated here were not generated only by the Vaca Muerta, but also by other source-rocks, of Jurassic and Cretaceous ages. The eastern boundary of the Neuquen Basin overlap by the Andean folds (Fold and Thrust Belt of Figure 22) is indicated. Courtesy R. Vially.*

The large deposits recently discovered between Cyprus, Israel, Palestine and Egypt seem to contain biogenic gas to a great extent.

Another mode of trapping is so-called methane hydrates or methane ice. In the language of chemists, they are clathrates. These are cages of water molecules that trap a molecule of methane, forming a solid that resembles ice, but which can be ignited (Figure 24).

They form under specific conditions of temperature and pressure. They can be found up to depths of a few hundred meters in what is known as Permafrost, a never-defreezing ground of continental arctic zones in Russia and Canada. It can also be found at sea in the sediments accumulated on the continental slopes, where the temperature, 4 °C, which is the temperature of the sea water at its maximum density, is somewhat higher than in permafrost, but at water depths higher than about 400 meters in the oceans and in the lakes, where the pressure becomes sufficient to ensure their stability at this temperature. Existing quantities are not known precisely, but they are large, some even claim much more than all other fossil fuels combined (*Makogon, 2010*).

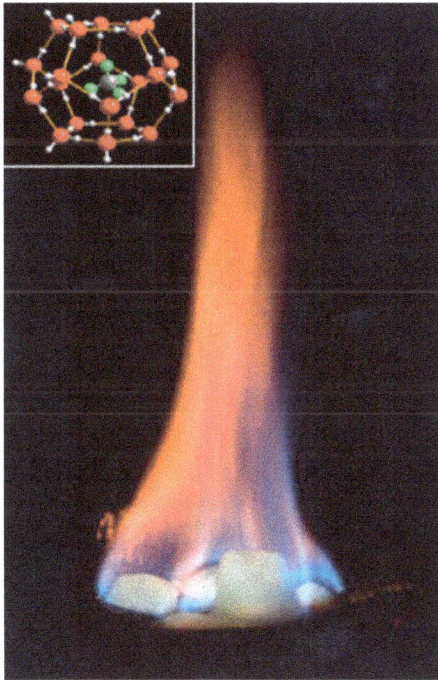

Figure 24 *Top-left representation of a clathrate element made up of a dodecahedron of water molecules (in red oxygen, white hydrogen) encasing a molecule of methane (gray carbon and green hydrogen). These cages are associated in more complex buildings. Below, a fire of blocks of clathrates. Source: CNRS in Wikipedia.*

This however does not make much sense: most of the gas hydrates are found in recent sediments deposited on the present continental margins, which date roughly of less than 100 million years, that is to say a small proportion of the geological history of fossil fuels, as it begins with the explosive development of living beings 2 to 3 billion years ago[14]. Moreover, since they are unstable, the ones we know have probably been generated very recently. It is therefore very doubtful whether they could represent a greater quantity than all known fossil fuels.

In fact, other studies (*Laherrère, 2009*) seem to show that their quantities in place would be of the order of the current reserves of the gas deposits. But quantities in place does not mean reserves. Only the profitable part can be considered as a reserve. However, they have a very large defect: they are very dispersed in the sediments in the form of discontinuous lenses on a small thickness and a very large surface. Nobody to date has managed to do better than small pilot plants without a future.

Japan, which is in great need of energy produced on its own territory, has made and continues to test offshore gas hydrates production, but apparently without much success. According to *Nelder (2013)*, after ten years and $ 700 million spent, about 1 million m^3 of methane was recovered from the Japanese Nankai site, while the same amount of imported liquefied natural gas purchased on the market would have cost about 50,000 dollars! The results are not better on the site of Mallick, in the permafost of the Beaufort Sea, Canada.

A recent review of gas hydrates is that of *R. Vially 2016.*

Gas hydrates are not necessarily formed from biogenic gas, but their ubiquity shows that this most often is probably the case: thermal gas could only come from a leak at the surface of a deeper gas deposit, inevitably very localized.

The durability of hydrates is linked to the maintenance of adequate temperatures and pressures. If there is no durable waterproof cover over the generating areas, the methane thus trapped is bound to dissipate over time as these conditions may change. This is one of the concerns of climate scientists, some of whom believe that global warming can lead to their release into the atmosphere. Methane is a much more powerful greenhouse gas than carbon dioxide. But if the environment changes, this methane could also be very quickly converted to carbon dioxide by aerobic bacteria before being released to the atmosphere. Cements of carbonates over sediments containing biogenic gas, but also over deposits containing thermal gas, whose origin would be methane thus transformed into carbon dioxide, have been observed, for example, in the North Sea (*Hovland and Irwin, 1992*).

It should be noted that gas hydrates could form in subsea gas pipelines that transport gas from offshore gas fields. Producers seek to avoid this formation because these

---

[14] Some oil fields are known to have formed about one billion years ago in Australia, Oman and Eastern Siberia, but also source-rocks are known to have produced oil 2 billion years ago, as in Oklo in Gabon! *J. Craig 2015* has recently produced numerous examples showing that the formation of oil and gas is as old as life itself, whose first indisputable signs of appearance date from about 3.8 billion years.

hydrates can block gas flow. For this, they use heated pipes or inject chemical inhibitors into the gas, in areas where this risk is high. These gas hydrates are also of concern to producers when offshore rigs are to be installed in areas rich in hydrates, because of the risk of sediment instability where they are present.

Another mode of trapping is by dissolving gas, in this case mostly thermal, in deep aquifers. We have seen that methane has a non-negligible solubility in water. A classic example is the deep aquifers of the Gulf Coast in Louisiana in the United States, where high-pressure aquifers from a few hundred meters deep contain up to 20 m^3 of methane under normal conditions by m^3 of water (*Durand, in Rojey et al., 1997*). This methane is sometimes recovered as a by-product of geothermal installations exploiting the heat of these aquifers. According to *Randolph (1977)*, the corresponding resources would be considerable on a world scale, much higher than those of hydrates. However, the reserves, that is to say the part that is recoverable profitably, are insignificant at the present time.

3.2 Solid fuels: coal and bituminous (oil) shale

3.2.1 From peat to anthracite: The formation of coal deposits

Higher plant debris can accumulate in large masses to form a sediment very rich in type 3 kerogen, called a peat, in which plant debris can be easily identified with the naked eye. The peat may form layers of plurimetric thickness and plurikilometric extension.

Large quantities of peat are currently being formed in the perpetually flooded swamps of cold areas of the northern hemisphere, from specific plants such as sphagnum and mosses, heather or sedges. These are the best known, which were used extensively for home heating and cooking, and even in the industry as heat source in countries like Ireland, northern Germany or The Netherlands some centuries ago. But their destiny will be essentially to remain in the state of peat, as they are largely installed on tectonically very stable zones which will not undergo significant burial for a very long time, called cratons or shields.

However, peatlands also form on the continent/ocean boundary of sedimentary basins undergoing deepening, in climatic zones with high plant productivity, and from very diverse plant associations. Examples are the deltas of the rivers of Indonesia, where debris of trees and other tropical woody plants accumulate. They are therefore progressively buried. They are covered by other sediments, over which other levels of peat are deposited, etc. So, what are called coal series, which may be several kilometers thick, are formed due to the permanence of the same tectonic and climatic situations in the same place for millions of years. The rocks constituting these series are generally alternations of sandstone and argillites, the latter often also containing kerogens of the same type (type 3), but in lower contents than in peat.

These peats thus subjected to increasing temperatures and pressures give rise to more and more compact coal, the kerogen of which has gradually been enriched with carbon, successively lignite[15] (brown coal), sub-bituminous coal, bituminous coal, anthracite (anthracite), and finally meta-anthracite (meta-anthracite).

This evolution is called coalification or coalification, or maturation, and the stage, called rank, is identified by the modification of the physical and chemical characteristics which accompany it: carbon content of kerogen, reflectance of vitrinite, parameters that increase with rank, volatile matter content, moisture content, decreasing... The ultimate ranks are only reached for burials of the order of ten kilometers or otherwise abnormally high geothermal gradients. Peat corresponds to the early diagenesis of Figure 11, while lignite corresponds to diagenesis, bituminous coal to catagenesis and anthracite to metagenesis. Meta-anthracite corresponds to the boundary of metamorphism.

The sketch of Figure 25 shows the successive stages of the transformation of peat into coals of increasing rank.

Bituminous coal (also known as hard coal) corresponds to the stage, which we have called catagenesis in the general evolution of kerogens during burial, that in which oil is formed and then wet gas. In some samples, fluorescent exudates of fine residual petroleum droplets, or impregnation of the immersion oil with hydrocarbons can be observed by reflection microscopy (Figure 5).

Yet geologists have long refused to admit that liquid hydrocarbons can form in coal series. In their minds, coal could only produce gas (firedamp). This is partly due to the fact that the reference coals were for a long time those exploited in Europe, of Carboniferous age. However, these have undergone tectonic inversions as shown in Figure 25. Possible deposits of petroleum formed in the Carboniferous would have disappeared by erosion. This is also due to the fact that the biomass from which these coals originated was relatively low in constituents that can generate liquid hydrocarbons.

On the other hand, more recent coal series, for example those from the Tertiary of Indonesia, have a much simpler geological history and many have not yet undergone significant tectonic inversion. Their kerogen, derived from tropical plants, contains in quantity liquid hydrocarbon-generating substances such as tree resins, and cuticular waxes which protected these plants from evaporation: there were much less in the plants generating European Carboniferous Coals. These series have given rise to oil deposits, sources of Indonesia's oil wealth (Figure 26).

We must not, therefore, speak of coal, but of coals, rocks of a great variety, according to their rank, but also of the peculiarities of the biomass of higher plants which gave birth to them. There is also a dimension linked to the importance of their mineral content, which is highly variable. It should also be noted that some coals, known as

---

[15] Lignite, which is a variety of coal, should not be confused as it is very often the case with lignin, which is an association of biological constituents of the supporting tissues (wood, roots, stalks, leaf veins, etc.) of the so-called higher plants (trees, herbaceous plants).

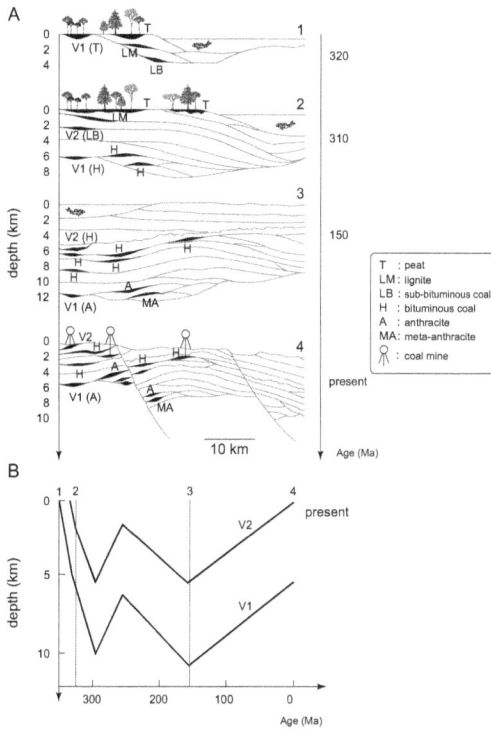

Figure 25 *Sketch of the progressive formation of a coal series. A: evolution of the basin and coal seams during geological time; B: history of burial over time of two coal seams (V1 and V2). The unit of time is the million years (Ma).*
Initial deposits of peat date from a little before 320 million years ago (Carboniferous). A first burial on about 20 million years results in the formation of a coal series. It is followed by a first tectonic inversion for about 30 million years, and an erosion of the upper part of the series. A new sedimentary basin, which deepens until about 150 million years ago, brings the oldest coals of the series to depths sufficient to transform them into meta-anthracite. Then a new tectonic inversion causes this set to go up and erode. Some of the coal levels are then sufficiently close to the surface to be exploited by underground mines. This requires depths of less than 1500 meters.

those found in India, known as Gondwana coals of the name of the super-continent where they were formed in the Carboniferous, and found especially in India, are very rich in inertinites for a reason poorly known, which decreases their calorific capacity.

The international trade classification, under the influence of the USA, is very re-ductive: peat, brown coal (soft coal), which is actually lignite, and hard coal, which combines sub-bituminous coal, bituminous coal and anthracite. Steam coal is the coal used to produce steam in boilers, mainly coal-fired power plants. It is mainly

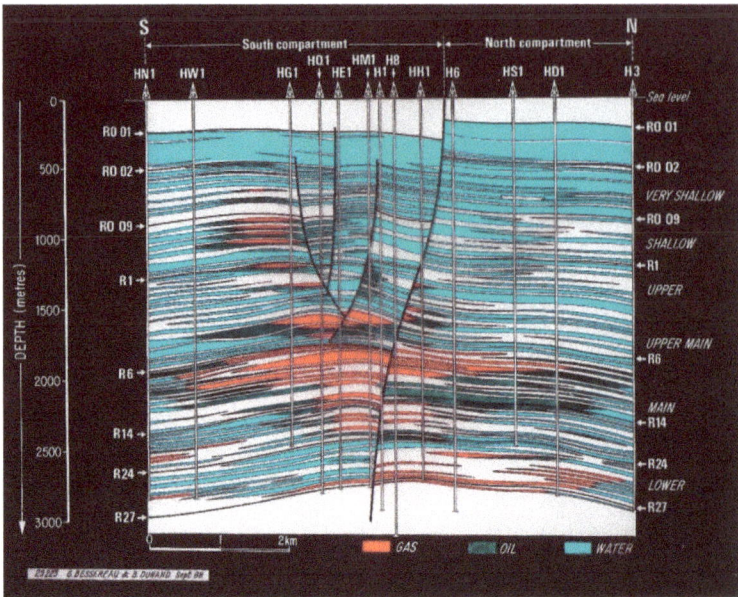

Figure 26   Cross-section of the Handil oil and gas field in the Mahakam Delta, Kaliman-
tan, Indonesia. Oil and gas occupy here sandy levels of a deltaic-type coal
series that began to develop there fifteen million years ago. They alternate
with clay sediments (white on the Figure) containing varying amounts of
kerogen derived from a biomass of tropical higher plants, rich in cuticular
waxes (i.e. contained in the cuticles of leaves) and in tree resins. Sandy levels
contain, in superposition, water (aquifer), blue, oil (petroleum), dark green,
and gas, in red, but sometimes only water. There are often coal seams in the
clay sediments, not shown here. This coal series, not this time associated
with oil and gas because they have not been trapped, is shallower on land,
and is exploited for its coal. The source of oil and gas is for this field the
coal series itself, but from a depth of about 3000 meters. The supply of
oil and gas to the sandstone levels which serve as a reservoir is made by
their dip downstream but also by the faults, which also serve as leakage
zones towards the surface. These levels are therefore temporary storage of
oil and gas, on a geological time scale, which will only last as long as the
inputs are greater than the leaks. This example shows both the complexity
of the architecture of a deltaic coal series and that of a so-called "multipay"
deposit, where there are many exploitable levels.

bituminous coal, possibly sub-bituminous, although lignite is also used to generate
electricity, particularly in Germany.

An excellent book to read in order to fully understand the very complex physico-
chemistry of coals is that of *van Krevelen (1993)*.

The International Energy Agency (IEA) makes the distinction between peat, lignite
(brown coal), sub-bituminous coal, coking coal, used to produce metallurgical coke,

Table 4 *Three different classifications of the marketed coals and main parameters used to characterize their rank. The heating value, here the lower heating value (LHV) indicates the amount of heat that can be produced by combustion (see Chapter 2.2 in second part). The values and limits given here are approximate because they depend on parameters such as the ash content or the nature of the plant biomass from which the coal is derived, which can vary substantially for the same rank.*

Europe	Peat	Lignite	Sub-bituminous	Bituminous high volatiles	Bituminous low volatiles	Anthracite
USA	Peat	Brown coal	Hard coal	Hard coal	Hard coal	Hard coal
Canada, France	Tourbe	Lignite	Flambant sec	Flambant gras	Gras	Anthracite
Rank	Low	Low	Medium	Medium	High	High
LHV (MJ/kg)	4–6	10–20	15–25	27–33	27–33	33–36
Humidity (%)	>50	25–50	14–25	5–10	5–10	1–6
Volatile matters (%)	>75	50	25–50	30–40	15–25	<10
Ash Content (%)	50	30–50	20–30	10–20	10–20	0–10
Vitrinite reflectance (%)	<0.3	0.3	0.35–0.45	0.5–1.25	1.5–2	2–5

which is a variety of bituminous coal richer in hydrogen than the average[16], other bituminous, excluding coking coal therefore, and corresponding roughly to what is otherwise called steam coal, and anthracite.

Although there are coal seams up to considerable depths, probably over 10,000 meters in some sedimentary basins, it is technically impossible to mine it at present if the depth is greater than about 1500 meters.

On the other hand, at the higher depths, it is sometimes possible to exploit the gas it contains (firedamp) by drilling of the oil type. It may also be possible one day to gasify the coal by underground gasification, as explained in Chapter 4.

3.2.2 *Bituminous (oil) shale deposits: rocks rich in kerogen, never having been sufficiently buried for producing oil*

Bituminous shales (oil shales) are sedimentary rocks rich in kerogen, but less so than coals. Their typical organic carbon content is about 10–15% of their dry weight.

---

[16] The hydrogen richness of coking coal makes it the coal with the highest calorific value. But also, this richness allows the appearance during the coking of a liquid phase, the mesophase, which then gives rise to a microporosity. The latter facilitates the circulation of air during the combustion of the coke in the blast furnace: The coke is very resistant mechanically, which prevents the compaction of the mixture of iron oxides and coke used to produce the cast iron.

Figure 27 *Oil shale outcropping in Morocco. We observe a "schistosity", in fact an alternation of lamines of variable mineralogical compositions according to the local conditions of sedimentation (see Figure 6), and a natural fracturation resulting from the mechanical stresses undergone by the rock during its geological history and which is revealed in during decompression during the gradual uplift to the outcrop. The average organic carbon content of such a rock is about 10% of its dry weight. Source: Wikipedia.*

Their kerogen is most often of marine plankton origin (type 2). However, the most important deposit, known as the Green River Shales in the United States, is of lacustrine origin (type 1).

These kerogens are rich in hydrogen (high H / C ratio) and therefore have a high oil generating potential. This, together with their richness in kerogen, explains their capacity to produce in quantity, on the order of 100 liters per tonne and more of shale oil by pyrolysis.

However, this is only possible because these rocks have never been buried deep enough to produce oil and gas. Indeed, if not, all or part of the petroleum potential of their kerogen would have disappeared, and they would no longer produce enough oil by pyrolysis for their exploitation to be profitable.

Still on questions of profitability, they are at present only exploited in quarries or in open pit exploitations.

The criteria defining bituminous (oil) shale are therefore both economic and geological.

The first bituminous shales to have been historically exploited for the industrial production of shale oil were the Schists of Autun, at Igornay in the Saône-et-Loire, France, in 1824. Some oil shales, such as the Kuckersite of Estonia, have long been used directly to produce heat or even electricity, as are coals.

The name bituminous (oil) shale does not seem very appropriate: they are rocks of a fairly great mineralogical variety, even though they are essentially clays and marls, and they contain no bitumen which is, as we have seen, a solid-state oil, but kerogen. The richest in kerogen are sometimes called sapropelites, of the term sapropelia, which means vase rich in organic matter in putrefaction. They also do not contain oil.

The quantities of bituminous shales so far recorded are considerable, and some estimate that the quantities of shale oil that could theoretically be produced by pyrolysis from them are much higher than the remaining reserves of oil. We are however very far from being able to switch from quantities in place to reserves, that is to say, quantities of oils produced in a profitable way!

The largest deposits, accounting for more than half of these quantities, are found in the United States: the Green River Shales formation, already cited, covering parts of the States of Colorado, Utah and of Wyoming.

Small exploitations in the Schistes-Cartons formation of the Lower Toarcian of Lorraine and Burgundy, as well as in the Schistes of Autun, still took place in France a little after the Second World War, and were resumed after the oil shocks of 1973 and 1979, but not for long. Many countries did the same with their bituminous shales resources.

The low net energy efficiency of shale oil production and the considerable environmental problems associated with it with current standards mean that for the time being, there are no longer any industrial exploitation in the world except for a small production in Brazil (Irati Shales). Old workings in Estonia (Kukersite) and China (Fushun Shale) are now arrested. The same applies to pilot plants in the United States (Shell Company's Mahogany Project in the Green River Shales) and Australia (Stuart Shales).

4

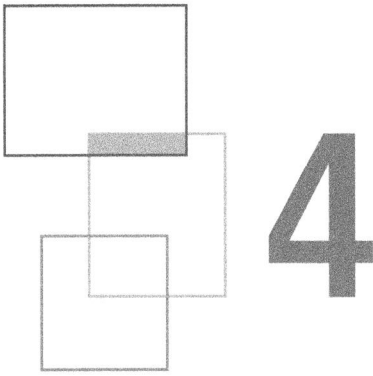

An overview of fossil fuels

So far, varieties of fossil fuels have been described separately. Now hereunder is a brief synthetic description of them:

Figure 28 is a geological section of a sedimentary ensemble containing the main categories of fossil fuel deposits that could in principle exist.

The history of this set is as follows: an ancient Palaeozoic (Primary Era) basement is initially stretched by tectonic forces and its left part collapses (at geological velocity), thus creating a set of moles and rift valleys (in geological terms horsts and grabens), which is gradually invaded by the waters. The funds are filled by conglomerates (gravels and pieces of rock packed in clay cement...), and sandstones (consolidated sands), resulting from the erosion of high points. Then, during the Mesozoic (Secondary Era), a sedimentary series is set up above, formed by successive strata of argillite (rock made up essentially of clay minerals) rich in kerogen of marine plankton origin (type 2), of sandstone, of argillite free of kerogen, and of a limestone formation (carbonates).

The tectonic forces then function in compression, which causes a tectonic inversion, that is to say an uplift of the sedimentary strata. The upper part of the series is then eroded as it rises. The compression causes an accentuation of the horst and graben structure of the basement (and a development of the faults that separate them), and a structuration in anticlines and synclines[17] of the sedimentary series above.

---

[17] For non-geologists: the sedimentary series are very frequently and more or less intensely folded under the effect of the tectonic forces. A fold forming a dome is an anticline, a fold forming a cuvette is a syncline.

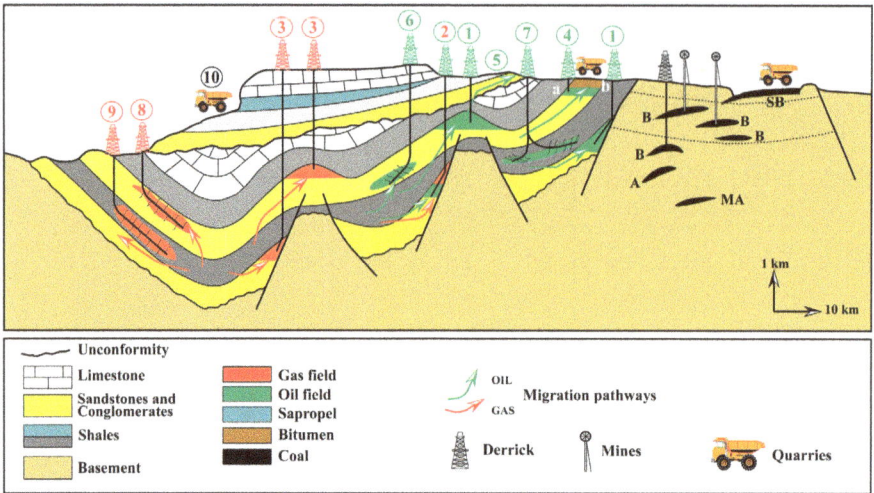

Figure 28 *Synthetic section combining the different categories of fossil fuels that can be found in sedimentary basins. 1: conventional oil; 2: conventional oil and gas; 3 conventional gas; 4a: extra-heavy oil; 4b: bitumen; 5: surface show; 6: tight oil; 7: shale (source-rock) oil; 8: tight gas; 9: shale (source-rock) gas; 10: bituminous (oil) shale (sapropelite). SB: subbituminous coal; B: bituminous coal; A: anthracite; MA: metaanthracite.*

The whole is then again progressively buried during the Cenozoic (Tertiary Era), and then settles on the surface of erosion (also called discordance or unconformity) a new sedimentary basin.

Finally, a new tectonic inversion brings out the upper part of this ensemble, which erodes to present its present physiognomy.

The rightmost part of this section is a non-collapsed part of the basement, the top of which has been eroded. It contains a coal series, containing therefore kerogen formed from higher plants (type 3), of Carboniferous age (from −360 to −300 million years). This series, because of its age and the erosion of its upper part, no longer comprises peat or lignite, which are the categories of coal less mature and therefore the newest and least buried of the coal series, but depending on the depth sub-bituminous coal (SB in the Figure), bituminous coal (B), anthracite (A) and meta-anthracite (MA).

A sub-bituminous coal seam near the surface is exploited in open-pit, that is to say in quarry, after stripping of the rocks surmounting the exploited seam: the limit of exploitability in open-pit, about 400 meters deep, is indicated in dotted lines. Bituminous coal seams are mined in underground mines down to the mining exploitability limit of about 1500 meters, also indicated by dotted lines. Below this limit (but also in the shallower coal seams but not exploitable in mines, for example if they are too thin), it is sometimes possible to extract by an oil-type well coal gas (Coal bed methane, CBM), a mixture of methane, carbon dioxide and sometimes

nitrogen, present in the porosity, joints and cracks of coal. This gas is what constitutes firedamp: it comes out from coal during the exploitation and when it invades
poorly ventilated underground mines, its inflammation is explosive and is the cause
of numerous deadly accidents. It was formed by the thermal degradation of the
kerogen of the coals during their burial, and part of it remained trapped in them.
This type of gas currently provides the United States with nearly 10% of its natural
gas supply, particularly through the San Juan Basin in New Mexico. Harvesting is
done from boreholes in the coal seams, directed from the surface to remain inside
these seams over long lengths. This is called horizontal drilling, although the veins
are rarely truly horizontal. The water in the formation must be pumped before the
gas can be pumped. Hydraulic fracturing of the coal seam must also be often carried
out beforehand. This consists of injecting water into the latter at a pressure sufficient
to exceed the mechanical strength of the rock, thus expanding the natural cracks or
creating new ones, which increases the permeability of the formation and allows the
extraction of sufficient volumes to be economically viable.

This gas, when it is released brutally during the operation of a mine, or slowly after
the closure of the mine, finally ends up in the atmosphere. The modern mines are
powerfully ventilated to avoid the danger of firedamp accumulations in the galleries.
Rather than dissipating it outside, this firedamp is increasingly recovered and used as
a source of energy for surface installations, or to produce electricity, with the added
benefit of burning methane, a powerful Greenhouse gas, while producing carbon
dioxide, much less powerful. This can be done in a number of ways: preventively, by
drilling in the coal seams during extraction, recovering the methane in the ventilation
air, or after the closure of the mine the gas which continues to come out. This set is
called coal mine gas (Coal Mine Methane, CMM).

It may be possible one day to exploit coal seams more deeply than the mining
limit, or coal seams too thin to be exploited otherwise, or coal seams at sea (In the
United Kingdom, for example) by the so-called underground gasification technique,
which consists of making a horizontal drilling between two vertical wells in the
seam to be exploited and then introducing a mixture of oxygen and steam at high
temperature through one of the wells. The combustion of the coal is thus caused in
oxygen deficiency, and produces so-called synthesis gas, a mixture composed mainly
of carbon monoxide (CO), hydrogen and methane, which is recovered by the second
well. The synthesis gas can be used directly in a gas turbine to produce electricity
or for heating, but also makes it possible to manufacture synthetic fuels or basic
molecules for petrochemicals, such as, for example, methanol (CH_3OH).

Underground gasification has already been used, for example in Uzbekistan at the
time of the Soviet Union, and has been the subject of some pilot installations,
including in France at Bruay-en-Artois. There is a research program in China, but it
has not yet resulted in commercial production.

This method is still very futuristic being too expensive, but also because its environmental dangers are not well understood. Coal formations are generally permeable
on a large scale, as they contain many sandstone bodies and are highly fractured

(see Figure 26). Thus, leakage of synthesis gas to the surface can be feared. There are also fears of important water rushes overflooding certain exploitations. On the other hand, coal can burn for a long time even in oxygen deficiency. It is currently the source of many fires, practically impossible to extinguish, of coal seams and mine quarters around the world. This type of accident could therefore occur at gasification sites.

Synthetic gas is, however, already produced from coal, but in plants using coal extracted from conventional mines, such as the Great Plains Synfuels Plant (GPSP) plant in North Dakota in the United States or those of SASOL Company in South Africa. China has an important program in this field, and rumor has even circulated that it is proposing to help Ukraine to establish such factories to enable it to avoid the import of gas from Russia.

In the left part of the figure, the kerogeneous argillite was, depending on the maximum depth reached, the source-rock for oil (oil zone) and then for an oil-gas mixture (wet gas zone), and finally dry gas. Oil and gases were collected by the sandstone levels (consolidated sands) located above and below the source-rock. Depending on the migration possibilities and the existing traps, accumulations have been created: oil (1) oil and gas (2) and gas (3). The arrows in the Figure 26 indicate the direction of movement (secondary migration) of the oil and gas from the source-rock to the traps.

A portion of the oil formed traveled upward in the formation of sandstone above the source-rock and was transformed therein when reaching the surface. In doing so, it gave birth to extra-heavy oils (4a) which remained trapped under bitumen (4b), produced by an even more pronounced alteration of the oil. This very viscous bitumen served as an impervious cover. We have tar sands in this location.

Through the fault passing through the oil deposit in the anticline trap, some oil passed through the clay cover, then the discordance and the sandstone level above it, to form an oil show in the form of a seepage of oil at the surface (5).

A part of the oil formed was also trapped in its upward movement in a very low permeable part of the sandstone formation (6) to form a tight oil field. Because of the very low permeability of the reservoir, it is necessary to recover this oil to make a "horizontal" drilling and then to perform multiple hydraulic fracturing (Figure 29).

All the oil was not expelled from the source-rock. The latter is extremely impermeable, its permeability is lower than that of a brick. In order to recover this oil (7), horizontal drilling and multiple hydraulic fracturing are also used in economically favorable sites. This oil is what the Americans initially called shale oil, an ambiguous term because it also refers to the shale oil obtained by industrial pyrolysis of bituminous shale (which would now be appropriate, as we have seen, to call kerogen oil rather than shale oil to avoid confusion). This situation is very similar to the previous situation, and the Americans now combine tight oil (6) and shale (source-rock) oil (7) under the same term light tight oil (LTO). The oil recoverable is indeed in both cases a light oil. The shale oil extracted from source-rocks also often comes not from the source-rock *sensu stricto*, which is the part that contains the kerogen having

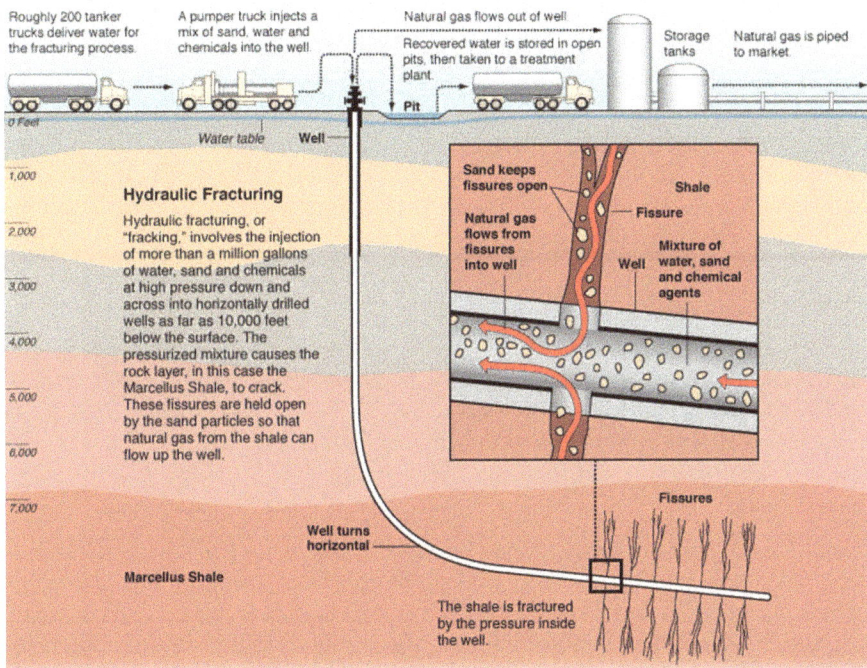

Roughly 200 tanker trucks deliver water for the fracturing process.

A pumper truck injects a mix of sand, water and chemicals into the well.

Natural gas flows out of well
Recovered water is stored in open pits, then taken to a treatment plant.

Storage tanks

Natural gas is piped to market

Pit

0 Feet

Water table   Well

1,000

Hydraulic Fracturing

Hydraulic fracturing, or "fracking," involves the injection of more than a million gallons of water, sand and chemicals at high pressure down and across into horizontally drilled wells as far as 10,000 feet below the surface. The pressurized mixture causes the rock layer, in this case the Marcellus Shale, to crack. These fissures are held open by the sand particles so that natural gas from the shale can flow up the well.

2,000

3,000

4,000

5,000

6,000

7,000

Sand keeps fissures open   Shale

Natural gas flows from fissures into well   Fissure

Mixture of water, sand and chemical agents   Well

Fissures

Well turns horizontal

Marcellus Shale

The shale is fractured by the pressure inside the well.

Graphic by Al Granberg

Figure 29	*Diagram illustrating the multiple hydraulic fracturing technique associated with horizontal drilling. The casing was drilled successively in several places to inject pressurized water and its additives. One notices the fineness of the cracks created: the rock does not therefore explode as some believe: the pre-existing cracks are enlarged and others created. The water injection pressure must be greater than the vertical stress on the sediment, about 60 MPa (600 bar) at 2500 meters depth as here. Source Wikipedia.*

produced the oil and / or gas, but in fact more permeable low thickness levels, such as, for example, dolomitic limestones or silts (very fine-grained sandstones), which are in fact tight reservoirs internal or in contact to the *sensu largo* source-rock, and which do not necessarily have a great continuity. It is also necessary for fracturing to be effective that the rocks be sufficiently brittle, which excludes very clayey zones.

On the gas side, gas is trapped in its upward movement in a very low permeable part (8) of the sandstone formation. This is called tight gas. It can be exploited by conventional drilling, and possibly by horizontal drilling and hydraulic fracturing to stimulate production where the permeability no longer allows the conventional exploitation.

The source-rock retained some of the gas it produced. In economically favorable conditions, it is also exploited by horizontal drilling and multiple hydraulic fracturing (9). The recovered gas is known as shale gas, actually source-rock gas. Again, situations 8 and 9 are very similar[18].

It is important to note that the yield from a well for the production of shale oil or gas, where horizontal drilling is associated with multiple hydraulic fracturing, contrarily to a production well of a conventional field declines very rapidly in the course of time, by about 50% in the first year and 80% in two years (Appendix 2). This requires constant drilling of a considerable number of wells to maintain the production of the area. On the other hand, while it takes five to ten years to begin to produce a conventional deposit, it takes only some months to start producing shale oil or gas.

Let us note in passing that shale gas and oil, contrary to what may be believed of their current popularity, have been known for a very long time. The small town of Fredonia in the State of New York was already fuelled by this type of gas in 1840! It was technological improvements and, above all, the period of high oil prices from 2005 to 2014 that allowed their massive development in the United States (see Second Part).

Finally, on a hillside (10) is an oil shale quarry. There is a sapropelite outcrop particularly rich in type 1 or 2 kerogen, which has not been buried deep enough to produce petroleum. The petroleum potential of the kerogen which it contains has therefore not been impaired. The argillite, which is the source rock of the oil and gas fields shown on Figure 28, may also, on its outcrop to the right of the Figure, be strictly exploited as oil shale because it has not been sufficiently buried and for that reason has not at this location significantly reduced its petroleum potential. It is of lesser interest than sapropelite because of a lower kerogen content. These oil shales will eventually be pyrolyzed in industrial plants to produce shale (kerogen) oil. Following the use of the same term for source-rock oil, this term has become ambiguous, and the Americans now also use as already said the term light tight oil (LTO) to refer to shale (source-rock) oil and kerogen oil to designate the oil produced by pyrolysis. The rock used to produce kerogen oil, i.e. oil (bituminous) shale, is also increasingly called kerogen shale rather than oil shale.

---

[18] It is important to realize that the shale (source-rock) gases have very varied compositions, such as the gases found in conventional reservoirs (see Table 2). The same is true of shale (source-rock) oils. The source-rocks also have various mineralogical compositions and mechanical properties. These parameters are never taken into account in non-professional speeches. They are, however, crucial in assessing the possibilities for exploitation. What interest, if the gas discovered is, for example, mostly nitrogen, what has happened in Poland?

There are many situations where the gas contains oil and the oil contains gas, as in conventional conventional deposits (see Figure 19). In the United States, shale gas operations are highly valued when it contains significant proportions of associated oil, given the high price of oil.

On the other hand, the high molecular weight liquid hydrocarbons, if any, are not extracted with the present techniques because of their greater affinity with the mineral surfaces and their higher viscosity. Source-rock oil is a light tight oil.

Second Part

Future prospects, climate and health risks

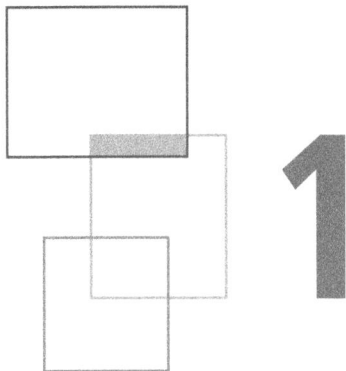

Some nomenclature

Natural conventional and unconventional oil and gas and synthetic oil and gas (synfuels and syngases)

1.1.1 *Natural oil and gas*

All the oils and gases listed in Figure 28 are natural products, in that they all originate in natural transformations in sedimentary basins.

Natural petroleum, which is almost of no use as it is, has to be refined into petroleum products. These are mainly, by increasing average molecular weight of the petroleum cut:

- naphta, i.e. a distillation cut consisting of liquid hydrocarbons of low molecular weight (number of carbons between 5 and 11), used essentially by the petrochemical industry – fuels, gasoline, kerosene[19] and diesel fuel, for land vehicles and aircrafts – domestic heating oil – heavy fuel oil for propulsion of ships or for the production of electricity. The distillation residues (the "bottom of the barrel") are optionally pyrolyzed to produce petroleum coke, which can be used as industrial fuel, but also as a carbon source for the manufacture of carbon electrodes, or even as an additive to coal for manufacturing metallurgical coke in coking plants (see Chapter 1.2). A by-product of refining of petroleum is hydrogen (H_2), which is used in the plant itself, in particular to remove if necessary sulfur from petroleum cuts,

---

[19] Not to confuse kerosene, also called jet fuel, which is the fuel of aircraft, with kerogen, which is as we have seen in part I the raw material for the formation of oil and gas in the sedimentary basins.

and is also exported to chemical plants. The proportions of these various products, for a standard refining protocol, depend on the composition of the initial petroleum, in particular the average molecular mass of its constituents.

There is also liquefied petroleum gas (LPG), which mainly contains propane and/or butanes liquefied under pressure. It is used mainly for heating and cooking, and secondarily as fuel for specially equipped vehicles. This LPG is produced largely from natural gas liquids, and more precisely, as seen in first part (Chapter 1.2), of the NGPLs recovered in the natural gas processing plants. Although its constituents are gaseous under normal conditions, the LPG thus produced is classified with oils in the production statistics. However, LPG is also produced, in about 40% of the total world production, in refineries from petroleum fractions. It then also contains light unsaturated hydrocarbons: ethylene, propylene and butenes.

As regards natural gas, it does not need to be refined since the gas shipped from gas plants is a dry gas, the non-hydrocarbon constituents (nitrogen, carbon dioxide, hydrogen sulphide, etc.) having been eliminated and non-methane hydrocarbons, recovered and accounted for with oil in the condensates and natural gas plant liquids (NGPL) categories (Chapter 1.2, first part and Table 5). It therefore contains practically only methane. It is mainly used for heating, industrial heat production, power generation, fertilizer manufacturing[20], or for propelling gas-powered cars, which use it in compressed form at 200 bars, this is called compressed natural gas (CNG).

The distinction is often made between conventional and non-conventional oil and gas:

Unconventional oils and gases originally referred to natural oils and gases that required more complex production techniques due to difficult access, deep offshore or polar zones, compared to conventional oils and gases, or unfavorable characteristics of the deposit, high viscosity, high depth and/or pressure, high hydrogen sulfide content, etc. This rather blurred boundary has varied over time depending on technological developments. Conventional oil has also been defined as that for which no other production stimulation techniques were used other than the so-called secondary recovery processes (i.e. water injection into the aquifer to maintain pressure), or as that of the deposits for which a water body could be defined, that is to say the existence of a clear contact between an aquifer and an oil or gas phase.

We will call here, as is now the most common use, conventional oils that can be produced from conventional drilling, even if deviated or even horizontal, but without systematically using heavy technology to stimulate production, such as hydraulic fracturing or SAGD (see below).

In Figure 28, the petroleum deposits corresponding to this category are those of types 1 and 2. The oil extracted from the types 6 and 7 deposits, which can only be produced by the use of hydraulic fracturing, is not conventional.

---

[20] These are mainly nitrogenous fertilizers such as ammonium nitrate and urea. The synthesis of their raw material, ammonia ($NH3$), from nitrogen requires hydrogen which is provided by methane. Methane also provides the energy needed for reactions.

| Table 5 | Summary of natural oil and gas varieties. Note that shale oil (7 on Figure 28 of first part) and tight oil (6 on Figure 28) are more and more combined in a single light tight oil (LTO) category. Indeed, the oil extracted from the source-rocks is actually also produced from tight reservoirs, which are located inside or in contact with the source-rock sensu stricto, that is to say levels containing kerogen. It is essentially shale oil, the tight oil being little represented. |

Natural oil	Natural gas
Conventional Oil	Conventional Gas
Extra-Heavy (XH):	Shale Gas
- Bitumen (Tar, Bitumen)	Tight Gas
- Extra-Heavy Oil	Coal vein gas (Coal Bed Methane, CBM)
Shale Oil	Coal mine gas (Coal Mine Methane, CMM)
Tight Oil	Methane hydrates
	Gas dissolved in deep aquifers

Extra-heavy oils (4a) and bitumen (4b) are transition categories. The world's major deposits are by far the bitumen (tar) found on surface and at shallow depths, less than 400 meters for current operations in Athabasca, Canada, and extra-heavy oils of the tar belt of the Orinoco (Faja Bituminosa) in Venezuela. The latter are classified in extra-heavy oils and not in bitumen (see Chapter 1): they are much less viscous than Canadian bitumen because, being on average deeper, between 350 and 700 meters deep and located in a country where the average ground temperature is much higher than in the Canadian North, the temperature of their reservoir is higher. They are also less altered by surface water. In Canada, exploitation is made by quarries or open-pit for depths of less than about 100 meters, or at higher depths by a doublet of horizontal drilling located one above the other, with injection of water steam at temperatures of around 350 °C in the upper borehole. Following the reduction in viscosity caused by its heating, the bitumen flows in the lower borehole. This is Steam Assisted Gravity Drainage (SAGD)[21]. Canadian bitumen is classified as unconventional oil.

Venezuelan extra-heavy oils are also produced by horizontal drilling, after injection of steam, but often also after injection of light hydrocarbons, to reduce their viscosity, and sometimes by SAGD. They will also be classified here in unconventional oils. It should also be noted that bitumen and extra-heavy oils require pre-refining at the site of their extraction, so that they can be transported to refineries (see chapter 1-1-2 below).

As for gas, conventional gas is the one produced from type 2 and 3 deposits. Unconventional gases are those extracted from type 8 and type 9 deposits, for which hydraulic fracturing is used. Gases produced from coal seams, coal seam gas (CBM) and coal mine gas (CMM), as well as gas hydrates and dissolved gas in deep aquifers are also classified in unconventional gases.

Table 5 summarizes all these categories of natural products.

---

[21] This downward flow is only possible because the density of the extra-heavy oils is slightly higher than that of the water.

1.1.2 *Synthetic oils (synfuels) and synthetic gases (syngases)*

The industry is capable of producing synthetic oils and synthetic gases, i.e. artificial liquid or gaseous substitutes for petroleum and natural gas. It is important to say a few words about them because in discussions on the future of global oil and gas production, the distinction is generally not made between natural oils and gases and artificial oils and gases. If this is of little importance to the user, according to the adage "don't look a gift horse in the mouth", this has a lot for future prospects, and it is very often a source of confusion in the discussions on the evolution of world oil production.

The artificial liquid products substituted for natural petroleum or its refined products are called synfuels: These are essentially at the moment, on the one hand biofuels (biofuels) also called agrofuels because they are derived from agricultural products, and on the other hand X-to-Liquids (XTL), produced from natural carbonaceous substances by complex industrial processes. X denotes the carbonaceous substance used as raw material.

Biofuels

These are biodiesel[22], mainly produced in Europe from triglycerides of vegetable oils (rapeseed, sunflower, imported palm oil, etc.), and ethanol[23] mainly produced in Brazil and the United States, from cane sugar in Brazil and maize starch in the United States.

There is often talk of producing biofuels from lipid-rich unicellular microscopic algae cultures. But the processes have not at the moment exceeded the stage of the pilot.

X-to-Liquids (XTL)

Gas-to-Liquids (GTL), i.e. fuels synthesized from natural gases by Fischer-Tropsch synthesis after their conversion to synthesis gas[24].

---

[22] Biodiesel, sometimes called diester, is produced by esterification of triglycerides, the main constituents of vegetable oils, with an alcohol, methanol (CH_3OH), produced by petrochemistry. It is therefore not entirely derived from biomass. Glycerin is produced at the same time.

[23] Ethanol (C_2H_5OH) is the alcohol produced by alcoholic fermentation of sugars. It is also produced from cereal starch, mainly maize. Starch, which is a polymer of glucose, is first transformed into glucose with enzymes.

[24] In order to produce synthesis gas, which consists essentially of carbon monoxide (CO) and hydrogen (H_2), from natural gas, coal, biomass or other carbonaceous substances, they are reacted at temperatures of the order of 1000 °C, with a suitable mixture of water steam and oxygen in the presence of a nickel catalyst. The Fischer-Tropsch synthesis, a method used to produce fuels, consists in reacting the hydrogen and carbon monoxide contained in the synthesis gas at 300 °C on an iron or cobalt catalyst. Thus, n-alkanes (paraffins) are formed, which are then isomerized, that is to say are converted in isomerization plants into isoalkanes (see Appendix 1), which are better suited to motors. One can also direct the reactions to make a whole range of products other than the fuels.

Coal-to-Liquids (CTL), i.e. fuels synthesized from coals by Fischer-Tropsch synthesis after their transformation into synthesis gas, or by the Bergius process[25].

Biomass-to-Liquids (BTL): these are fuels produced by converting biomass into synthesis gas and then Fischer-Tropsch synthesis. For this purpose, the lignocellulosic part of the plants may be used, much more abundant than vegetable oils and sugars used to make biofuels. Despite the hopes that these raise, it is only at present time in use as pilot productions.

Shale-to-liquids (STL): these are the shale oils (kerogen oils), produced by pyrolysis at temperatures of the order of 500 °C of so-called bituminous (oil) shale (Category 10 in Figure 28), of which we have seen in the first part I that in reality contained no bitumen, but immature kerogen. This shale oil, which has nothing to do with shale oil from source-rocks, is a simile crude oil much richer in hetero-elements (see Chapter 1) than natural crude oil, and thus more difficult to refine. For the time being, this is an anecdotal category, its production being insignificant on a global scale, for lack of profitability. Nevertheless, significant quantities were produced at the time, from the mid-19th century to the mid-20th century, for example in France (see Chapter 3.2.2.1).

It should be noted that GTL and CTL are essentially substitutes for petroleum-based fuels, the utility of which is evident only insofar as the supply of petroleum-based fuels is no longer sufficient to supply demand. But the counterpart would be a faster decline in gas or coal reserves, to the detriment of other uses of these, such as electricity generation.

Extra-heavy oils and bitumen (4a and 4b) are classified by some in the XTL category. Indeed, as in the case of STL, pre-refining in a facility called upgrader to make them an "oil" usable by a conventional refinery is necessary.

Syngas is a gas produced by various industrial processes: gases produced by blast furnaces and cast-iron converters into steel (carbon monoxide and a little hydrogen), or by coking plants (hydrogen and methane), synthesis gas (carbon monoxide, hydrogen, methane) that would not be used for syntheses of fuels or chemicals, produced from natural gas, naphta, biomass or coal (or which would be produced by underground gasification of coal). These gases can be used directly to produce heat and electricity. They were widely used until shortly after the Second World War, before the development of natural gas production.

Table 6 makes the list of current synthetic fuels (synfuels) and synthetic gases (syngases).

Synfuels production currently accounts for about 3% of the so called "all liquids", also said "total oil", production (see next chapter), and consists mainly of biofuels (agrofuels) derived from plant sugars and triglycerides. Ethanol from Brazil and the United States accounts for the bulk of biofuels. Europe produces mostly biodiesel.

---

[25] The Bergius process consists in reacting together coal and hydrogen at a few hundred degrees under pressure, to produce a synthetic oil which must then be refined.

| Table 6 | *Synthetic fuels and gases varieties.* |

Synthetic fuel (synfuel)	Synthetic gas (syngas)
Biofuels *Biodiesel* *Ethanol* XTL *Gas-to-Liquids (GTL)* *Coal-to-Liquids (CTL)* *Biomass-to-Liquids (BTL)* *Shale-to-Liquids (STL) (shale oil, kerogen oil)*	Gas produced by blast furnaces, converters, or coking plants Syngas produced from natural gas, naphta, biomass or coal Gas produced by underground gasification of coal

Syngas production represents less than 1% of "total gas" production.

Energy Agencies, the most known of them being the International Energy Agency (IEA) (www.iea.org) based in Paris, and the United States Energy Information Administration (EIA) of the US Department of Energy (DoE), based in Washington (www.eia.gov), have the daunting task of presenting all these natural and artificial products in a simplified form to the attention of public authorities, industry and opinion.

1.1.3 *The barrel of oil, what does it contain?*

International accounting of oil production and of oil consumption (i.e. refined products) is done, under the influence of American culture, in barrels. This is a unit of volume, not of mass or energy (we shall return to this point), which is worth 159 liters. As far as oil production is concerned (we do not deal with consumption in this book), the barrel known to economists, the media and, therefore, public opinion is the barrel of "all liquids" (also called total oil, oil supply), that is to say it contains all the categories of natural oils and artificial oils which are listed above. The unit most used in international production statistics is the million barrels per day (Mb/d, d denoting here *dies* and not day). 1 Mb/d represents an annual output somewhat less than 50 million tonnes (see Chapter 2.2).

Figure 30 and Table 7 show the value and composition of world liquid oil production in 2014, in Mb/d, according to the Energy Information Administration (EIA) of the US Department of Energy (DoE).

It is observed that the barrel of all liquids petroleum contains:

- two categories which come not from oil but from gas: condensates and NGPLs. As seen in Part I in Chapter 1-2, and in Table 5, condensates are liquid hydrocarbons recovered at the head of wells from gas associated to oil fields, or from gas fields, while NGPLs are liquid hydrocarbons or liquefied gases still contained in the gas which are subsequently recovered in the gas processing plants. Note that the unconventional shale (source-rock) gases, denoted 9 in Figure 28 and the tight gases, denoted 8 in Figure 28, may also contain condensates and NGPLs, which

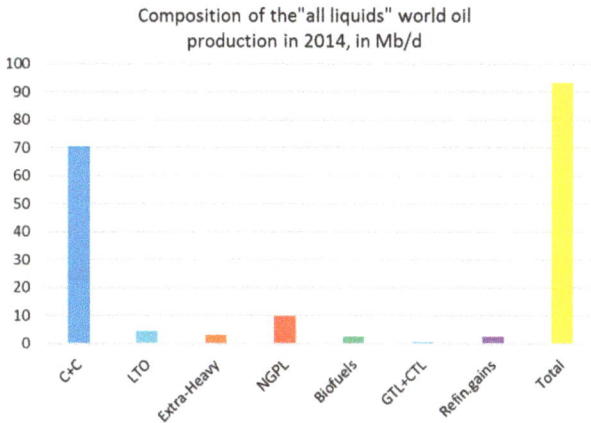

Composition of the "all liquids" world oil production in 2014, in Mb/d

Figure 30 *Histogram of the composition of world all liquids petroleum production in 2014, according to the United States Energy Information Administration (EIA).*

Table 7 *Composition of world production of all liquids oil in 2014 according to the EIA, in Mb/d and in percentage.*

Component	Mb/d	Percentage
Conventional crudes + condensates (C+C)	70.3	75.4
Light Tight Oils (LTO)	4.4	4.7
Extra-Heavy	3.2	3.4
NGPLs	9.8	10.5
Biofuels	2.5	2.7
GTL+CTL	0.5	0.6
Refinery gains	2.5	2.7
Total all liquids	93.2	100

are accounted for in LTO for the condensates and in the global NGPL category for the NGPLs. On the other hand, unconventional gases from coal (CBM, CMM) do not contain condensates nor NGPLs;

- synthetic oils (synfuels): biofuels, GTL, CTL…;

- a very special category, the refinery gains: These represent the volume gain between the refinery entry and exit. It is due to the fact that the refining products have a greater total volumic mass than that of the initial charge introduced into the refinery because they are hydrogenated in the refinery. Volume accounting (barrel) has the perverse effect of classifying in oil-producing countries, countries that simply refine it such as the Netherlands Antilles!

A very large part of the barrel, about 86% of the total worldwide, consists of the categories C + C (conventional crude + condensates) plus NGPLs. The contribution of the conventional crude is impossible to estimate precisely because the producers almost systematically mix it with condensates and account for this mix. The condensates could at the moment make up 7 to 8% of the C + C assembly.

In spite of the great media attention devoted to them, the proportions in the barrel of all liquids of non-conventional oils (LTO and Extra-heavy) represent only 8.1% and synfuels (biofuels, GTL, CTL) 3.3%, totaling 11.4%. However, NGPL's extracted from oil and gas shales might represent an additional 3%.

Each component has its own production dynamics: it is therefore necessary to examine for each of these production dynamics in order to analyze the future possibilities of producing all liquids petroleum (see Chapter 3-2-5).

It should be noted at the outset that the way in which these categories are grouped varies from one agency to another in their most common publications, which is a constant source of confusion for commentators and requires some inconvenient gymnastics to compare data. Regarding the two main ones, EIA and IEA, if they combine conventional oil with condensates (C + C), the IEA separately counts the LTO and the Extra-heavy (XH), while the EIA, in its most common publications, agglomerates them with C + C to make a single category "crude oil" or "petroleum". The IEA often includes all synfuels (biofuels, GTL, CTL…) under the same label, while the EIA details them.

1.2 Categories of coals and their use

Let us again insist that there is not one but coals, from peat to meta-anthracite, and that the variety is as great as that for oils and gases. Considering them as a single substance, as is generally done in the media and in many statistics, is extremely reductive.

However, the current uses of coal are fairly limited. In decreasing order of importance, the main uses are electricity production, coke production from coking coal for iron metallurgy (manufacture of pig iron and steel), and the production of heat for industry.

Electricity generation, which currently consumes about 70% by mass of the world's coal production (and represents just over 40% of the world's electricity production), and industrial heat production, which consumes about 10%, can use all coal varieties, grouped for these uses under the "steam coal" catch-all label. Coals having a high calorific value (see Chapter 2.2) are of course preferred, and therefore have the greatest commercial value. For the manufacture of electricity, however, as in Germany for example, lignite is used, despite its low calorific value. For reasons of profitability, it is then necessary to install the plants as close as possible to the production areas, in order not to bear the cost of transport.

Coking coal, about 15% by mass of the world consumption, belongs to the so-called bituminous coal category (see first part, Chapter 3.2.1). The latter correspond, in the successive stages of evolution of the coals, to the stage called catagenesis in the formation of petroleum. Coking coals are in this category those which are the richest in hydrogen. They correspond to the beginning of catagenesis when the kerogen of the coals lost most of its oxygen while still retaining much of its hydrogen (see Figure 32 below).

Coke is manufactured in coking plants, where the coal is pyrolysed in coke ovens at temperatures of up to about 1300 °C. A wide variety of valuable liquid and gaseous effluents are produced. Coking is essential to eliminate a number of elements that could hinder the manufacture of cast iron, sulfur for example. The coking coal also has the peculiarity to turn into a microporous solid very resistant mechanically, coke. Coke is used to reduce the iron oxides constituting the iron ores: the microporosity allows a good circulation of the air and therefore a good combustion, and the mechanical resistance allows the good mechanical strength of the charge in which the coke is mixed with the ore. These three factors: elimination of harmful elements, microporosity and mechanical strength are the reasons for the success of the coke. Various attempts to manufacture coal coke have been made in Europe since the early 17th century. The development of an effective process by Abraham Darby in England in 1709 revolutionized iron metallurgy, which previously used charcoal, and was one of the key elements in the development of the Industrial Revolution in Europe. It is however known that coal coke was produced in China several centuries ago, but this did not lead to an industrial revolution.

The development of a microporosity is precisely due to the hydrogen richness of the coking coals and more precisely of hydrogen present in the form of aliphatic chains as described in Chapter 2.3. The pyrolysis initially causes the creation in the coal of a semi-liquid phase, called mesophase, in the form of microdroplets disseminated in the mass. The coal then becomes pasty. The subsequent volatilization of this mesophase leaves behind this microporosity, the paste then transforming in a highly resistant microporous solid.

These peculiarities of coking coal make them the coals with the highest commercial value. It is also the most difficult variety of coal to replace in its uses.

However, it is now known to produce coke from coal mixtures which are less hydrogen-rich by adding small quantities of heavy fuel oil, or petroleum coke, a carbonaceous residue of petroleum refining.

Domestic heating, which represented a major use of coal until the 1960s, now accounts for only a few percent of world production. The same applies to the production of chemicals (carbo-chemistry).

The use of coals to produce fuels (CTL), as the Germans did during the Second World War, and South Africa during Apartheid and even now with the Sasol company (whose production of CTL is of 10 Mt per year), or to produce gas, is still anecdotal at the global scale. The Chinese company Shenhua produced about 1 million tonnes per year of CTL by Bergius process in Erdos, Mongolia in recent years and is reportedly planning to significantly expand this production, but information on this is rare. It should be noted that these are highly energy consuming and highly polluting processes, which are not compatible with Western environmental standards.

2

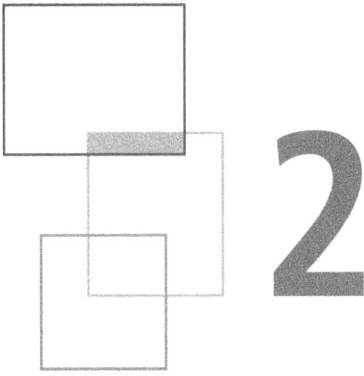

Some quantification

To what depths are deposits of fossil fuels found?

As exploitable deposits of fossil fuels can be formed only in sedimentary basins, these depths can *de facto* only be less than the thickness of the sedimentary series: few, even stacked one on the other by the play of successive burials and tectonic inversions, have cumulative thicknesses in a basin greater than 10 km, the world record being apparently that of the basin of the South of the Caspian Sea, whose maximum depth would be of a little more than 20 km.

2.1.1 *Oils and gases*

Figure 31 shows the distribution of reserves (see below for a discussion of the meaning of this term) of oil as a function of the depth of the discovery well relative to the level of the ground or the seabed for the non-US world and Canada. The statistics, due to J. Laherrère, cover about 14,000 conventional deposits on land (onshore) and at sea (offshore).

It is observed that half of the total volume of reserves was discovered at less than 2500 meters deep and that very few were discovered at depths greater than 5000 meters.

In fact, the current depth of these deposits is for the majority less than the maximum depth they have reached during their geological history, as sedimentary basins have virtually all undergone one or more tectonic inversions and therefore sediment erosion when they were brought to the surface. However, it rarely plays on more than a few hundred meters, and that does not change the picture.

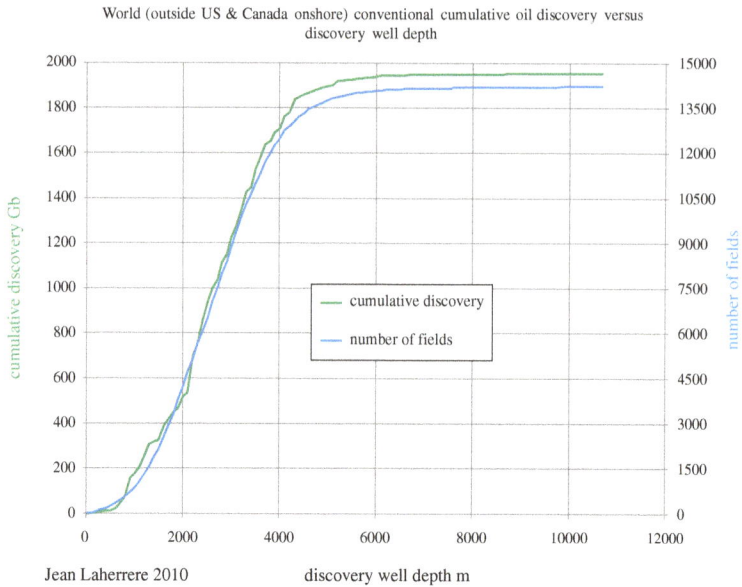

World (outside US & Canada onshore) conventional cumulative oil discovery versus discovery well depth

Jean Laherrere 2010

Figure 31 *Distribution of depths and discovery volumes in the world in 2010, excluding onshore deposits in the United States and Canada. For onshore in the United States and Canada, the number of wells drilled by small independent producers for which data has not been tabulated is far too large for statistics to make sense.*

It may be objected that this distribution is due to the technical limitations of drilling processes. It would come from the fact that the cost of drilling increases much faster than the depth of the boreholes, so that very little deep boreholes are done, and the small deposits found at these depths are not taken into account because of their exploitation cost.

On the other hand, with the exception of very soft rocks and/or low geothermal gradient, it is not possible to effectively drill well beyond 10,000 meters, the performance of drilling tools decreasing with depth and especially with temperature! The world record, 12,260 meters long, has long been a well drilled in the Kola peninsula in Russia, which was only recently beaten by a well of 12,370 meters drilled at Sakhalin by the company Exxon Mobil.

Oil and gas, due to their overall upward movement during their migration, generally accumulate above their formation zone. However, the latter can hardly reach more than 6 to 7000 meters of maximum depth with respect to the ground or to the sea bed, the kerogen of the source-rock having then practically exhausted its potential for formation of hydrocarbons unless the geothermal gradient is exceptionally low, which occurs, for example, in sedimentary basins with very thick salt strata, such as the Brazilian offshore basins.

The distribution of gas deposits as a function of depth is not very different from that of oil deposits. If it is formed at depths that are on average higher than oil. However, gas is also more mobile in the sediments and its deposits may therefore be even higher above its formation zone.

The deposits located at sea are at depths similar to those of the deposits on land, if one considers only the sedimentary series which contains them, and thus taking as reference the bottom of the sea. If reference is the sea level, then add the water slice. The thickness of the latter is very variable, up to 3 km in the case, for example, of certain deposits exploited in the Gulf of Mexico.

This thickness of water slice will undoubtedly be exceeded for the deposits that are planned to be exploited off the coast of Brazil, which are found in the so-called subsalt, or offshore the coast of Angola.

2.1.2 *Coals*

The coal series can have considerable thicknesses, are rarely more than 5 to 6 km. However, they can be buried at much greater depths, and then can be found in the zone of metamorphism, where important mineralogical transformations occur. This is how anthracites and meta-anthracites are formed. There are, however, technical limitations to the possible depth of operations. The limit for open pit exploitations is about 400 meters. For underground mines, it is about 1500 meters: this is due to the temperature, which makes the human presence very difficult even with strong ventilation, but even more to an operating cost that grows much faster with the depth, and with energy expenditure exceeding recoverable energy in the coal extracted. At higher depths, it is possible to drill for the gas contained in the coal seams (CBM) and it may become possible to exploit them by underground gasification. The depth limitations will then be roughly those of the oil wells.

2.1.3 *Bituminous (oil) shales*

Since the bituminous (oil) shales could not have been significantly buried (otherwise its kerogen would have lost its potential in pyrolysis oil because it had already produced oil and gas), they are *de facto* at shallow depth, generally lower than a km. On the other hand, it is currently only known to exploit them in quarry or open pit, which limits their depth of exploitation to a few hundred meters. Underground gasification might be possible, but it is *a priori* much more difficult than for coals because their organic carbon content is lower. There are very complex projects of underground pyrolysis, mainly in the United States to exploit the formation of the Green River Shales, where the world's largest quantities of oil shale are found. However, these projects have not as yet been successful: after 30 years Shell abandoned in 2012 the Mahogany *in situ* project in Colorado, consisting of electric heating in wells to pyrolyse schists and thus produce *in situ* shale oil (kerogen oil, STL), with freezing all around to prevent water coming.

2.2 The calorific value of fuels, the source of their economic interest

The economic advantage of fuels lies chiefly in their ability to produce heat by reaction with oxygen (combustion), in other words in their calorific value. This calorific value depends initially on their carbon, hydrogen and oxygen content. If the combustion is total, the carbon will produce carbon dioxide and the hydrogen will produce water, with the release of heat. The oxygen already present in the fuel decreases the calorific value. The position of the fuels on a diagram such as that in Figure 32 allows a rapid visualization of the C, H and O composition of most of the fuels and fuels listed here.

The calorific value is the amount of heat produced by the combustion of a unit of fuel mass. It is necessary to distinguish the higher heating value (HHV) from the lower heating value (LHV). Both are expressed in joules/kilogram.

The HHV is equal to the LHV plus the heat generated when the water vapor produced by the combustion of the hydrogen contained in the fuel condenses.

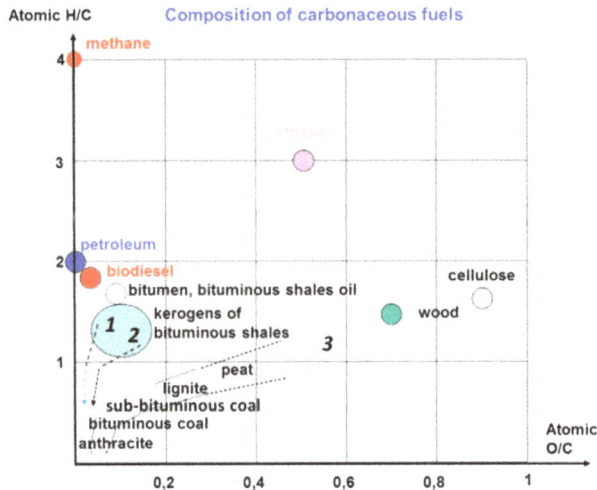

Figure 32   *Position of the main fossil fuels in a van Krevelen-Durand diagram (rectangular H/C versus O/C diagram, in atomic ratios) and comparison with carbonaceous fuels, as well as wood and cellulose. 1, 2 and 3, position of the average pathways of types 1, 2 and 3 of kerogens. For the definition of kerogen types see first part, Chapter 2.*

Hydrogen has a HHV of about 141 MJ/kg and an LHV of about 121 MJ/kg. For pure carbon (in graphite form), since it does not contain hydrogen, HHV and LHV are of course identical, about 33 MJ/kg.

For fossil fuels, the gap between HHV and LHV ranges from about 2% to about 10% of the LHV depending on the hydrogen content[26], with the greatest gap for the hydrogen-rich fuel, methane, which has a LHV of about 50 MJ/kg and a HHV of 55.5 MJ/kg, 11% more than the LHV.

To evaluate the global energy potential of fossil fuels, one unit, the tonne-oil equivalent (toe) is used, which is in principle the calorific value of one tonne of oil. However, the calorific value of oils is quite variable according to their composition (Table 8). Thus, the energy equivalent of toe was arbitrarily set at 10,000 thermies. Thermie is worth 1 million calories. The calorie, defined in 1824 by Nicolas Clement, is a unit older than the joule. Long used to measure the quantities of heat, its value has fluctuated slightly over time, from one definition to another and according to the perimeter of applications, between 4182 and 4204 joules. In short, for the International Energy Agency (IEA) the toe is 41.86 billion joules (GJ). The value retained by the French Administration was for a long time 42 GJ!

Table 8   *Approximate LHV of fuels shown in Figure 30. The range of values for coals and even for petroleum reflects their variety of origins and compositions.*

Combustible	Cellulose	Wood (10% of humidity)	Lignite (Brown Coal)	Subbituminous coal	Bituminous coal.	Anthracite	Ethanol	Biodiesel Bitumen	Crude oil	Methane
LHV (MJ/kg)	14	16	10-20	15-25	27-33	33-36	29,7	36-38	40-44	50

A great difficulty in making quantitative assessments of the energy content of world oil production is that the quantities of oil produced are evaluated by volume and not by mass. It is much easier to continuously measure a volume than a mass at the outlet of a well. Moreover, under the influence of the dominant culture in the oil world, American culture, the unit of volume used is the barrel, a unit which is worth very nearly 159 liters and is not recognized by the International System of Units (SI).

---

[26] Hydrogen is, as we have seen, a major constituent of fossil fuels and its content determines, to a large extent, their calorific capacity and therefore their interest and industrial uses. By mass, hydrogen content is 25% for methane (80% in atoms). It is 17% for butane (about 71% in atoms), 14% for gasoline and diesel (about 66% in atoms), 11% for heavy fuel oil (about 64 % in atoms), and 10% for bitumen (about 60% atoms). The LHV is the most commonly used for fuel qualification of combustion energy contained (Table 8). Indeed, the heat which could be recovered by condensation of the water vapor produced is not present in the usual uses (it is so in the so-called condensing boilers). The LHV of the marketed fuels varies greatly depending on their nature, but also on a large number of other parameters, in particular their content of water or minerals, as well as the mode of packaging.

Due to differences in density from one oil to another, conversion to tonnes is difficult. An old rule of thumb is 7.3 barrels per ton on average. As a result of the increasing contribution of relatively low-density products to the world's supply of "all liquids" oil, this ratio was close to 7.7 in 2013, according to the BP Statistical Review, which gives statistics in both barrels and tonnes. A commonly used production unit is one million barrels per day (Mb/d or Mb/d, d meaning dies and not day), or 365 million barrels per year. If you take a conversion factor of 7.3 or 7.7, that is 50 or 47.4 million tonnes per year.

On the other hand, the calorific value of conventional petroleum per unit mass, but also that of unconventional petroleum and other products that are accounted for by the agencies in oil production (Table 7) is variable.

To account for these variability, the publication of statistics in tonne-oil equivalent (toe), which is a unit of energy[27], rather than in barrel, which are not so far commonly done by Agencies, would be desirable. In the first analysis, however, it can be assumed that at the scale of world production, 1 tonne of oil statistically has an energy content close to 1 toe.

The uncertainty of production statistics (see Chapter 3.2) adds to these difficulties.

As a result of this, it is not clear what the energy content of the world's all liquids production is. If the production in barrels is still increasing, the energy content of an average barrel has been decreasing considerably for some time due to the increasing contribution of low-energy categories and the energy available in this production does not seem to increase any more. The amount of energy usable per inhabitant of the planet, given the increase in the world population, but also the decrease in the rate of energy return on energy invested (see below) is certainly decreasing.

Xavier Chavanne (2015a) has recently made an interesting proposal: to express the production of oil in barrel-oil equivalent (boe). This time the boe is a HHV, worth 6.12 GJ. This proposition starts from the empirical observation that the HHV of a conventional barrel of oil or condensate is on average 1 boe. On the other hand, it is relatively easy to convert the energy contained in a barrel of NGPL, knowing its composition, into boe. This method would hardly change the language habits of the profession, but would make it possible to follow much more exactly the energy content of world production.

When oil replaced coal as the world's largest source of energy, which occurred around 1965, the tonne-oil equivalent (toe) replaced the tonne-coal equivalent (tce) in the Statistics. Constructed in the same way as the toe, the tce is worth 7000 thermies,

---

[27] The unit of energy in the International System of Units (SI) is the joule. But it is an extremely small unit compared to global energy consumption. We saw that a toe "is worth" for the International Energy Agency 41.86 billion joules (GJ)! Worldwide primary energy consumption is currently about 13 billion toe (Gtoe). 1 Gtoe is "Worth" $41.86| \times |10^{18}$ joules (EJ)! We also use the Wh, which is worth 3600 joules, and its multiples, kWh (kilo, 1000 Wh), MWh (mega, 1 million Wh), GWh (giga, 1 billion Wh), TWh (tera, 1 trillion Wh). 1 Wh is "worth" 3600 Joules, 1 toe is worth 11.55 MWh, and 1 Gtoe is worth 11.55 billion MWh.

and thus 0.7 toe, or 29.3 GJ. This conversion coefficient is still used by many to make an assessment of the energy content of world coal production, which is estimated in tonnes. But the energy content of coals is much more variable than that of oils (Table 8). It varies greatly according to their rank and the nature of the plant biomass which gave rise to them, but also according to the water and ash content. According to the BP Statistical Review, a tonne of coal with the average composition of current world production is currently worth only about 0.5 toe (0.7 tce) and not 0.7 toe (1 tce). Most commentators, who usually reason in tonnes and not in toe, or when they reason in toe, use this coefficient of 0.7 rather than 0.5, thus overestimate the energy content of the world's coal reserves. In China, by far the largest producer in the world at present, the average energy content of coal produced has been declining for some time, from 0.52 toe/t in 2005 to 0.45 toe/t in 2012 (D. Fridley, personal communication).

One must also be aware that the production of fossil fuels requires energy. To estimate the energy actually available to humans, it has to be deduced from the energy content of the production the amount of energy used for this production. This is the notion of Energy Return on Energy Invested (EROEI, or more simply EROI) that is quantified by a ratio that of energy available in production over energy used to produce it. An EROI of 1 at the production level therefore means that there is as much energy produced as the energy used to produce it, and that exploitation therefore no longer makes sense from an energy point of view. We have seen, for example, that this was the case for coal extracted by mines greater than about 1500 meters in depth.

It may still have an economic value if it produces an energy better valued than that used to produce it. Global society, which works globally in a closed system, cannot anyway use more energy to produce fuels than the energy available in these fuels. An EROI of 1 worldwide, including all forms of energy and including food, would mean its end.

This fundamental aspect of energy production was for a long time totally neglected, probably because the EROIs have long remained very high.

EROIs estimates are increasing in number at this time but are still very fragmented and uncertain, given the difficulties of having sufficient direct data. There is also disagreement as to the calculation methods and the choice of the "perimeter" of the calculation: should the calculation of the EROI be limited to the energy balance of the extraction of the raw material, or should it be extended to the providing of products marketed to the consumer[28], or to 'useful' energy, that is to say that which actually serves to provide the desired services, or even to the energy consumed by

---

[28] The notion of EROI at the consumer level goes beyond the notion of final energy, that is to say the fraction of the energy initially available in the natural source (its primary energy) which is marketed to the consumer. Let us take the example of Canadian bitumen: these bitumen have in their deposit a lower heating value (LHV) that can be estimated at about 38 MJ/kg. A certain amount of energy is needed to extract, refine and convey these refined products to the consumer. The final energy is what remains of their initial energy when it reaches the consumer; but in this particular case, gas is used to produce the high-temperature steam required for extraction, that is to say, another source of energy than the bitumen itself.

the material and economic environment which allows them to be used (for example, the energy needed to manufacture roads used by motorists)?

At the production level, it is observed that the EROI is now decreasing rapidly for oil and gas (Figure 33), as deposits require more and more energy to find them and to exploit them. Some deposits already have very low energy balances. According to *Chavanne (2015b)*, if the energy losses from the crude oil still in the field to the consumer of refined products would currently be in the order of 11% for conventional oil (of which only 2.5% for production) i.e. an EROI at the consumer level of 9, it would be about 40% for the Canadian oil sands (of which only 25% for production), which corresponds to an EROi of 2.5 at the consumer level!

Global Oil and Gas EROI Values and Trends (1990-2010)

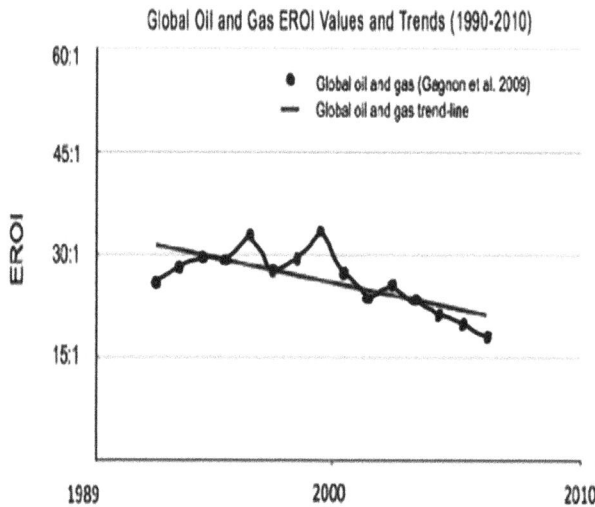

Figure 33 *Evolution of the EROI for the global oil and gas world production from 1992 to 2006, according to Gagnon et al. (2009).*

For coal, the EROI at the extraction level would still be very high. Commonly cited values range from about 50 to about 80. However, this could well be exaggerated: According to D. Fridley (personal communication), the EROI at the production level of Chinese coal, which accounts for roughly half of world production, would now be down to 17!

Among recent research on EROI, in *Hall et al. (2014)*, a detailed study of the EROIs of fossil fuels, as well as other sources of primary energy, can be found for different perimeters. *Weissbach et al. (2013)* made study on EROIs of electricity generation by energy sources used, and *Aucott and Melillo (2013)*, made a study on EROIs for shale gas production. This research shows the complexity of the subject but also a lack of homogeneity in the methods followed by the various authors and their results. However, they also show that we cannot avoid thinking about this in order to identify the future of fossil fuels.

3

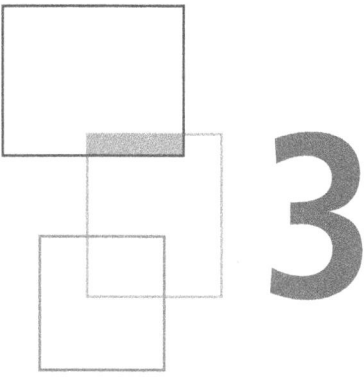

Future prospects of fossil fuels

3.1 A very difficult problem: the evaluation of fossil fuel reserves

3.1.1 Reserves, what is it?

The quantities that can be extracted with the technological know-how and in the conditions of economic profitability of the moment are so-called reserves, it is fossil fuels like any mineral raw material, iron, copper, gold, uranium...

It is important to distinguish between the reserves and the quantities in place in the deposits. The latter represent the quantities existing in the deposit, but they can never be extracted in their entirety.

Reserves and quantities in place can never be precisely known, because they can only be estimated by statistical methods: Now these can by nature only provide a probability of existence. In fact, the reserves of a deposit cannot be known until it is definitively abandoned, and the quantities in place will never be precisely known.

At the scale of a deposit, the ultimate reserves (in brief Ultimate) are its cumulative production from the beginning of its exploitation until its abandonment. Globally, they are the ones that have been extracted from all the deposits whose exploitation has been definitively stopped, together with those that can be extracted from the deposits still in operation, deposits discovered but not yet exploited, and deposits still to be discovered. This notion therefore has a more speculative character than the previous notions. Indeed, it is difficult to predict not only the discovery of deposits, but also the evolutions of technology and the economy.

Even more speculative is the notion of resources, which includes the quantities in place in known or discovered deposits, and even adds accumulations that are currently unusable or unknown for the moment, all of which might become accessible through hypothetical technological revolutions and at a date also hypothetical. Ultimately, each particle of organic substance contained in the earth's crust can be considered as a resource of fossil fuel. However, in order for it to become a reserve, it is necessary to find a technology that enables it to be extracted profitably and by using less energy than it can provide. The same applies for instance to methane in the atmosphere (about 5 billion tons).

To clarify the concept of reserves and to show its uncertainties, let us first consider the way in which reserves are evaluated in practice on the example of a conventional oil deposit (Figure 34).

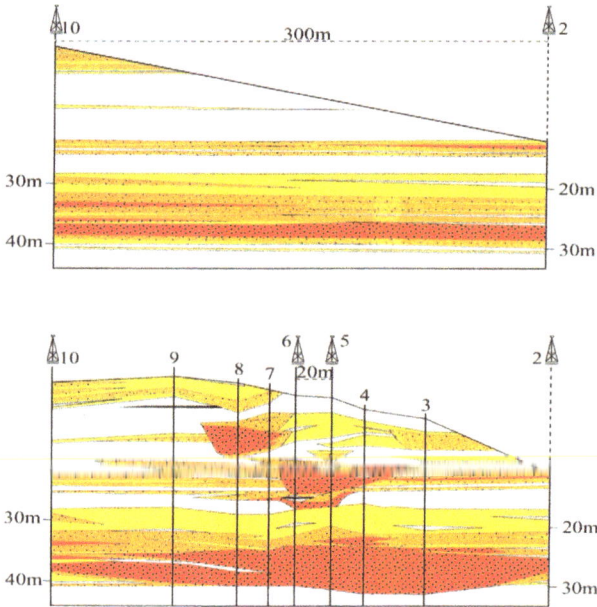

Figure 34   *Modifications of the geological model after drilling delineation wells. In red the zones supposed to contain the oil. Courtesy Christian Ravenne.*

The first step is that of discovery by one so-called discovery hole. With only 2 boreholes (top of Figure 34), we can only construct a simplistic model of the geology of the deposit and its production possibilities, from the information collected in these boreholes: the nature of the rocks traversed, the permeability of the reservoir formations (those containing the petroleum), flow tests, composition and physical properties of oil, etc. Uncertainty here is far too great to make an operating decision and some additional delineation wells are drilled. It can be seen how the information gathered on these new wells in this case modified deeply the initial geological model.

Geological models are based on geologists' knowledge of the types and structures of sediment deposits over time, as well as information provided by geophysical studies which realize a tomography of the structure, by the analysis of the logs, i.e. records of the physical properties of the rocks made during drilling, and by the flows, composition and physical properties of the collected fluids which provide information on the permeabilities, and therefore on the possible extraction rates. They are refined using geostatistical methods. It must be understood that the new model, even though more probable than the first one, is still nevertheless uncertain.

At this point, the decision must be made to exploit the deposit or abandon it. We start by estimating the quantities in place, that is, the total quantity of oil present in the deposit, based on the data collected, which is interpolated by geostatistical methods, and therefore sometimes with very large uncertainties. For a conventional deposit such as the one shown above, the ratio of the recoverable quantities to the quantities in place, this is called the recovery rate, varies greatly from one deposit to another, from 5% to 80%. But it is on average less than 30% for the moment (according to J. Laherrère, out of 17200 fields making up the bulk of the world's reserves: median rate according to the number = 25%, median rate according to the volume of the reserves = 41%, mean = 26%), despite the improvements made over time by the so-called enhanced recovery techniques. This is due to the limitations inherent in the physical mechanisms of the flows of polyphase fluids in porous media (see first part, Chapter 3.1.2). But its calculation is also necessarily tainted by strong uncertainties, for if we will know in fine the quantities extracted, we will never know precisely the quantities in place. However, an *a priori* assessment of this recovery rate will be necessary, depending on the geologists 'and producers' experience of the type of deposit discovered.

A production model for the deposit is then developed, based on the available information, which evaluates the possible flow of oil over time according to the infrastructure.

This model is then confronted with an economic model of price evolution over time, which makes it possible to evaluate the profitability of production.

From all this is drawn an estimate of the cumulative production over time and therefore of the reserves, that is to say of the quantities which can finally be extracted.

Because this estimate is made with statistical methods, it can by construction only give probabilities.

Proved reserves, or 1P reserves, are those volumes that are estimated to be more than 90% probable of being able to extract them profitably over the life of the exploitation: This is a low estimate of actual reserves.

We also estimate probable reserves, in fact proved + probable, called 2P reserves: these are those that we have more than 50% probability of being able to extract.

Then, we estimate the possible reserves, in fact proved + probable + possible (3P) reserves, i.e. those that we have more than 10% probability of being able to extract: It is a high estimate of the recoverable reserves.

The best statistical estimate of the ultimate reserves for the average of the deposits that have been exploited so far has shown *postmortem* to be that of the 2P reserves, and it is generally this datum that guides the company's decision. And thus also that of the banks, when they accept to take part of the risk, to provide the financing that the company may ask them for[29].

For someone not accustomed to these gymnastics, it must be realized that they are not exact quantities, but quantities with a certain probability of being finally extracted, and that the margin of uncertainty over their value is high. On the other hand, these quantities are all the smaller as the probability of being able to extract them is high.

During the operation, new wells will be drilled, the deposit will be better known, and the reserves 1P and 2P will be reassessed on a regular basis. In reality, an increase in the reserves 1P does not increase the ultimate reserves of the reservoir by the same amount, that is to say the total quantities which can ultimately be extracted from them. Simply it put reserves from "probable" to proved reserves.

If we want to know the ultimate quantities which can be extracted from a deposit, and the remaining quantities of these ultimate quantities, it is therefore the knowledge of reserves 2P, also called technical reserves, and of their re-evaluation, that it is necessary and not that of 1P reserves.

However, 2P reserves are generally not published by oil companies, which are not even allowed to do so in the case of companies listed on the New York Stock Exchange. The media and even the agencies in charge of energy statistics have no direct access to it, except for the few countries in which they are the subject of public government statistics or producer associations: The United States for the Gulf of Mexico with the Bureau of Ocean Energy Management (BOEM), Canada until 2009 with the Canadian Association of Petroleum Producers (CAPP), the United Kingdom with the Department of Energy and Climate Change, (DECC) and Norway with the Norwegian Petroleum Directorate (NPD) for the North Sea. However, they can be evaluated on the basis of documents available in oil companies, collated by consulting companies such as Information Handling Services (IHS), Rystad or Wood and Mackenzie. It is however a patient specialist work.

On the other hand, it is essential, if one wishes to have a correct idea of their evolution over time, to report all the revaluations of reserves 2P to the year of the discovery of the deposit, and not to the year of the revaluation. This is called back dating (retroactive revaluation).

---

[29] Regarding the financing of the exploitation of a deposit: more and more it is practiced on the principle of a production sharing contract between a State which is the owner of the deposit and a company or a consortium of companies that operates it, and that is what is called the operator. Banks lend to the State and not to the operator. The latter then becomes, by its production, the guarantor of this claim!

The approach for conventional gas is strictly identical to that described for conventional oil. The average recovery rate is much higher here than in the case of petroleum, with about 80%, because natural gas has much less affinity for rocks surfaces than oil, and is much less viscous than oil.

With respect to unconventional natural oil and gas, the bulk of those is currently Canadian bitumen, Venezuela's extra-heavy oils, and shale (source-rock) oils and gases from the United States and Canada.

For bitumen and extra-heavy oils, the reserve estimate cannot be as elaborate as for conventional oil, since it depends much more on the methods of exploitation, which are variable. These methods of exploitation are also much more burdensome to implement than for conventional oil, and their possible development is therefore slower. This is particularly true of Canadian bitumens, much of which is mined in open pit.

For shale (source-rock) oil and gas, the method used for conventional oil and gas simply does not make sense (insert).

Calculation of reserves of shale (source-rock) oil and gas

From the example of Figure 34, it is seen that the first step in the calculation of reserves is the construction of a geological model of the reservoir rocks containing the oil and gas: precise localization, dimensions, continuity, distribution of porosities and permeabilities... This is based on knowledge on sedimentology of sediment types, geophysical studies, petrophysical data and permeabilities estimated on rocks sampled into the wells available, and interpolated by geostatistical methods. Nothing or almost none of this is possible for source-rocks. We have seen moreover that the oil and gas extracted from the source- rocks often come not from the source-rock *sensu stricto*, which is the part which contains the kerogen having them produced, but in fact from more permeable levels of small thickness, such as, for example, dolomitic limestones or silts (very fine-grained sandstones), which are in fact tight reservoirs internal to the *sensu largo* source-rock, or situated in contact with it. These levels do not necessarily have a great continuity. It is also necessary for fracturing to be effective that the rocks are brittle, which excludes very clayey zones.

Knowledge of oil and/or gas compositions is also of great importance in assessing actual operating possibilities. In the United States, the exploitation of shale (source-rock) gas is in many cases cost-effective only because it contains associated natural gas liquids (condensates and NGPL). In contrast, the discovery of a source-rock gas that would be very rich in carbon dioxide or nitrogen would not be of interest.

It is not currently possible to adequately describe the fine structure of the source-rocks or the extent of their favorable levels and their precise mechanical characteristics in order to make reliable estimates with a few wells as is done for conventional deposits.

Since it is necessary to get an idea of the extractable quantities before deciding to invest large sums on an exploitation, we proceed in an empirical way after a pilot phase where the average productivity of the wells is observed. It is also known from experience (see Appendix 2) that the production of a well declines rapidly, about 80% to 90% of initial production in three years. It is therefore possible, having observed the decline curves during this pilot phase to evaluate the extractable average quantities during the first 3 years as a function of the number of wells drilled.

From experience, it becomes also possible to delimit what are called sweet spots, that is, areas where the wells are more productive than the average in the exploited formation, and where new wells will be concentrated.

Every exploitation is thus a special case, and many wells have to be drilled before having a good idea of the potential of a formation. Outrageous statements about the production potential of shale oils and gases on a global scale are fanciful and will remain so until there is sufficient drilling in areas considered promising. The example of Poland, originally announced as having the most abundant reserves of shale gas in Europe, and of which most of the oil companies have now withdrawn, should give cause for reflection. There is therefore a great need to develop a methodology for direct reserves estimation based on a much better knowledge of the sedimentology and mechanical characteristics of the source-rocks, the prediction of the composition of the oil and/or gas they contain, as well as by geophysical methods.

It should be added that the production of shale oils and gases, and consequently the valuation of their reserves, is much more sensitive to market prices than conventional oil and gas. Indeed, their accumulations are more diffuse than those of the conventional ones, in the sense that they are more poorly delimited and cover larger rock volumes with lower average concentrations. The cost of the investments to produce them is therefore significantly higher, for the same quantity produced, than for conventional oil and gas.

For coal, proved reserves are those quantities which have a high probability of being recovered under current economic and technological conditions. This is roughly equivalent to 1P oil or gas reserves. The term ambiguous of proved resources refers to those whose quantities and quality have been established by reliable geological data and supported by analyzes. So, there are no reserves at all, in the absence of a development plan and an economic calculation. Rather, this notion is similar to that of quantities in place used by the petroleum industry. We will see that in the case of coal, there is too much tendency to identify these so-called proven resources with reserves, which leads to overestimating the real possibilities.

The same is true for oil shale, whose resources are considered immense, while there are nearly no reserves, given the absence of profitable operations at present, or for gas hydrates, declared as has been seen by some as the future of humanity, but of which there is no presently any profitable production.

3.1.2 *What reserves can be expected in the future?*

As a preamble, let us recall once again that reserves are not measured quantities, but quantities which have been estimated by statistical methods, with a high degree of uncertainty. These quantities are all the smaller that their probability of existence is the higher. Giving reserve values without giving the corresponding probability and, as far as possible, the margins of uncertainties, makes little sense. But this is never done in current publications!

We will only give here broad outlines on a global scale, without detailing country by country. We will see that, in fact, it is impossible to get a clear idea from the data provided by the Energy Agencies, which are the only ones to be disseminated by the media.

3.1.2.1 The oils

Let us first recall that, for oil extracted from conventional deposits, the 2P reserves (probability of existence greater than 50%) were the best statistical estimate of the ultimate reserves, in brief ultimate, that is to say total quantities which can be extracted during the period of exploitation of the deposit.

Figure 35 shows the evolution since 1920 of the remaining 2P reserves of extractable oil from conventional deposits for two estimates: one was published in 1998 by *Colin Campbell and Jean Laherrère,* in Scientific American (The End of Cheap Oil), the other is a reassessment made by J. Laherrère in 2016. The two curves are close to one another and have a maximum in 1980: This means that since that date, the Ultimate of oil discovered of the conventional deposits no longer offset the volumes of oil consumed, despite price increases and technological advances increasing the recovery of oil in the deposits in operation! The proportion for the period 2003–2013 was about 1 to 2 (average annual 15 Gb found against 30 Gb consumed)!

Incidentally, recall that about 90% of the 2P reserves of conventional oil are found in only 10% of the fields, as shown in Figure 36 where the fields are classified by decreasing 2P reserves. This statistics is made for 2010 from a panel of 21,600 conventional oil fields worldwide containing 2200 Gb of 2P reserves of crude + condensate, however excluding US + Canada said non-frontier, i.e.[30] including only Alaska and offshore fields.

The world's conventional oil production thus depends essentially on a small number of very large deposits, called giant and supergiant. For the most part, they were discovered between 1950 and 1980, and many of them are now close to exhaustion.

How is it then that the media-economic discourse constantly refers to growing reserves? And why does it consistently repeat that there are still reserves for 40 years, which are assumed to be at constant consumption (which in itself makes no sense,

---

[30] The reason for the exclusion of the United States and Canada outside of Alaska and offshore is that the oil was exploited from the very beginning by a very large number of producers, most of them very small, and the statistics are therefore impossible. As a result of their long oil history, there would have been in operation in the United States in 1988 about 40,000 oil fields and about 35,000 Gas fields. This number has undoubtedly decreased, but in 2012 there were still about 560,000 oil producing wells and 480,000 gas-producing wells!

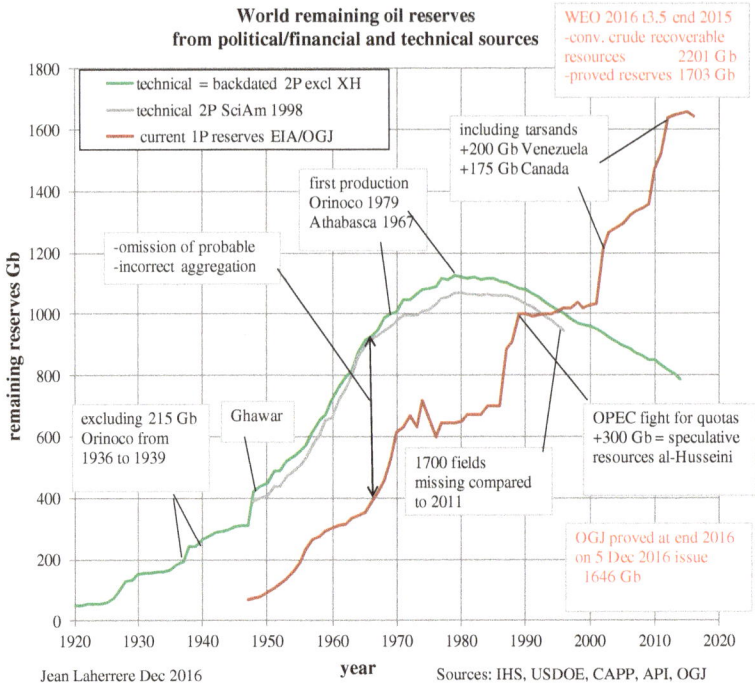

World remaining oil reserves from political/financial and technical sources

WEO 2016 t3.5 end 2015
-conv. crude recoverable
resources 2201 G b
-proved reserves 1703 Gb

- technical = backdated 2P excl XH
- technical 2P SciAm 1998
- current 1P reserves EIA/OGJ

including tarsands
+200 Gb Venezuela
+175 Gb Canada

first production
Orinoco 1979
Athabasca 1967

-omission of probable
-incorrect aggregation

excluding 215 Gb
Orinoco from
1936 to 1939

Ghawar

1700 fields
missing compared
to 2011

OPEC fight for quotas
+300 Gb = speculative
resources al-Husseini

OGJ proved at end 2016
on 5 Dec 2016 issue
1646 Gb

remaining reserves Gb

year

Jean Laherrere Dec 2016   Sources: IHS, USDOE, CAPP, API, OGJ

Figure 35 *Estimates of oil reserves 2P & 1P, in Gb, from 1920 to now: grey: 2P reserves curve established by C. Campbell and J. Laherrère in 1998; green: re-evaluation by J. Laherrère in December 2016; brown: average of estimates by various agencies of reserves 1P. It should be noted that the 2P reserves reported here concern only conventional oil, therefore without extra-heavys (bitumen and extra heavy oils), while 1P reserves include those. LTOs are also not considered here. But their importance was negligible before 2010, date of the beginning of their rapid growth in the United States. Also, it is not possible to evaluate correctly their reserves. It should also be noted that the addition of 1P reserves from different fields is statistically false (Capen, 1996). This in fact leads to an underestimation of these reserves, as indicated in this Figure. This is nevertheless the common practice of agencies. On the other hand, the rules of the SEC prohibiting the publication of reserves 2P, it is impossible except as here by a patient reconstitution from the data available in the companies to compare the evolution of the remaining 2P reserves to that of the 1P (declared reserves).*

because in any case we will not consume a constant quantity of oil for 40 years, and not at all the following year!)?

Simply because this speech refers to the reserves declared by the companies, or collated by the energy agencies. These reserves, announced as proved reserves, and which should therefore be the reserves 1P, are in fact not, for they are in fact calculated by methods in which technical criteria are mixed with politico-economic criteria which render them poorly credible. If the financial markets control Agencies like

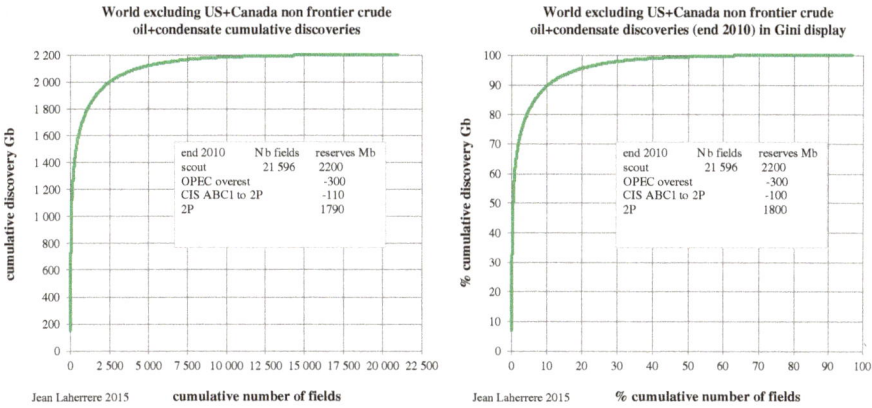

World excluding US+Canada non frontier crude
oil+condensate cumulative discoveries

end 2010	N b fields	reserves Mb
scout	21 596	2200
OPEC overest		-300
CIS ABC1 to 2P		-110
2P		1790

World excluding US+Canada non frontier crude
oil+condensate discoveries (end 2010) in Gini display

end 2010	N b fields	reserves Mb
scout	21 596	2200
OPEC overest		-300
CIS ABC1 to 2P		-100
2P		1800

Jean Laherrere 2015   cumulative number of fields

Jean Laherrere 2015   % cumulative number of fields

Figure 36 *On the left, reserves 2P cumulated according to the number of fields (fields) for a population of almost 22,000 oil fields excluding extra-heavy oil. On the right,+ the same data represented in a Gini diagram, where the discovered reserves are expressed as % of the total discovered and the number of fields as a percentage of the total fields. 90% of the 2P reserves are in 10% of the fields and 60% in less than 1% of the fields (184). The 2P reserves of the largest, Ghawar, discovered in 1948 are estimated at 150 Gb. The number of giants (> 500 Mb) would be 540 with 59 supergiants (> 5000 Mb). Courtesy Jean Laherrère.*

the Securities and Exchange Commission (SEC) of the United States are likely to be aware of financial risks, they do not know much about geology: They have established their own rules for the calculation of reserves to be declared by oil companies in their balance sheet, which have changed over time, and that very long have not made much sense. Adds to this confusion the development of production sharing contracts (note 26), which make it more difficult to distinguish between yours and mine, and vary the declarations of reserves of the companies according to fluctuations in oil prices.

Moreover, there are also several classifications of these so-called proven reserves: for the United States, and consequently all the companies listed on the New York Mercantile Stock Exchange (Nymex), in particular all major western companies (majors), the financially-inspired classification of the Securities and Exchange Commission (SEC) as said above, but also that of the Society of Petroleum Engineers (SPE), which is more technical. In Russia, there is the ABC1 classification. OPEC has its own classification, rather of a political nature, as we shall see.

There is above all the almost systematic absence of back dating, the reassessments of "proved" reserves being attributed to the year of the revaluation and not to the discovery year of the deposit, giving the impression of a reserve growth, in fact an artifact, but appreciated by the companies that show themselves in a favorable light for investors. In the United States, there is also the fact that many "independent" small producers report proved reserves according to a rule of thumb, multiplying to calculate their reserves their production of the year by a factor that is always the same, about 10, year after year. On the other hand, proved reserves are published by some analysts on the basis of a survey of producers even before the necessary studies have been made! The example of the US Lower 48 of the United States, i.e. outside Alaska and Hawaii, for conventional oil (Figure 37) shows the inability to use the ratio of so-called proved (current) reserves to production to predict the future.

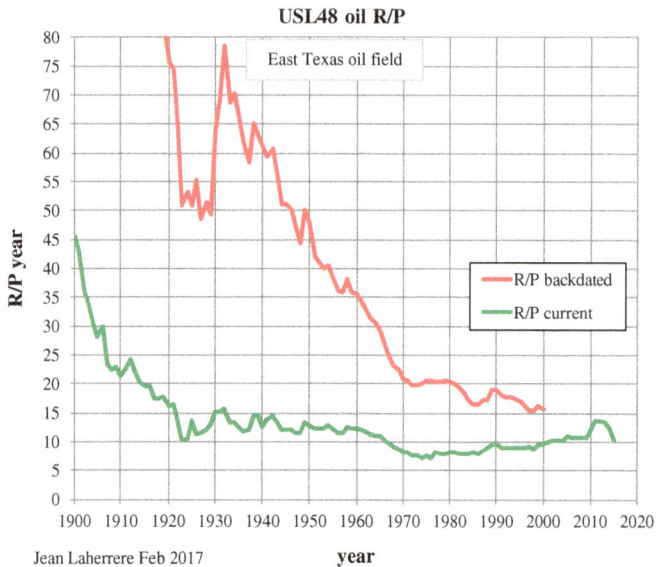

Figure 37 *Comparison of reserves to production ratio (R/P) for reserves 2P backdated to the year of discovery of the deposit and for the declared proved reserves (current), for the oil production of US 48, before the take-off of shale oil, that is to say essentially for conventional oil.*

Table 9 shows the very "financial manipulation" and unreliability of certain declarations of reserves: According to Bloomberg, this is the difference between the reserves reported to the SEC and the "resources" presented to investors by oil and shale gas companies in the United States. This picture is all the more surprising since, as has been said, there is no reliable technique for estimating reserves for these productions!

These declared "proved" reserves are also subject to political manipulation. A classic example is the addition of 300 Gb to the reserves declared by the countries of the Organization of Petroleum Exporting Countries (OPEC) from 1986 to 1989, after the oil countershock of 1986 and the decline in production guided by Saudi Arabia. According to a senior official of the Saudi state-owned company, Aramco, this re-evaluation came at the time of a bidding up between OPEC member countries to obtain better production quotas. These reserves had no physical reality. Despite the absence of new discoveries and significant production, they have even continued to increase. Iran and Iraq are reviving this bump from time to time.

Table 9	*"Shale" oil and gas in the United States: reserves declared to the Security and Exchange Commission (SEC) and "resources" announced to investors, in millions and billions of barrels. Source: Laherrère according to Bloomberg.*

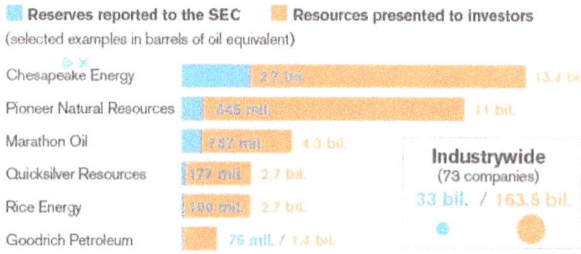

Reserves reported to the SEC   **Resources presented to investors**
(selected examples in barrels of oil equivalent)

Chesapeake Energy	2.7 bil.	13.4 bil.
Pioneer Natural Resources	645 mil.	11 bil.
Marathon Oil	1.07 mil. / 4.3 bil.	
Quicksilver Resources	177 mil. / 2.7 bil.	
Rice Energy	190 mil. / 2.7 bil.	
Goodrich Petroleum	76 mil. / 1.4 bil.	

Industrywide
(73 companies)
33 bil. / 163.5 bil.

Source: Company presentations and SEC filings   **Bloomberg** Visual Data

The declared "proved" reserves have also increased by the more or less arbitrary introduction of 200 Gb of Venezuela's extra-heavy oils and 175 Gb of bitumen from Canada. Again, these values are dubious, because even if enormous quantities are in place and are relatively well indexed, their recovery rate is largely speculative. In particular, it depends very much on the type of exploitation. For bitumen in Canada, about 80% for open pit operations, it is only 10% to 20% per SAGD. For Venezuela's extra-heavy oils, it is only 8% by cold extraction, compared with 25% for extraction with steam injection. In Venezuela, it also depends very much on the geopolitical context: the rate of recovery has declined after the departure of foreign companies.

On the other hand, these are oils whose dynamics of development are slower than those of conventional deposits, and is similar to that of coal production. If one wants to try to predict the future of oil production, it is better not to mix these reserves with those of conventional oil, because, as will be seen later, the size of the tap is more important for the future than the volume of the barrel.

The remaining proved reserves also vary from one agency to another (Figure 38), although they feed on the same sources.

If in Figure 35 the 300 Gb of speculative OPEC reserves are subtracted, and the almost 400 Gb of announced reserves of extra-oils as well, so as to make the comparison only for conventional oil, the remaining declared "proved" reserves would be nevertheless, after the countershock of 1986, almost continuously increasing and would intersect a little before 2010 those of the remaining 2P reserves. This is impossible by definition: the remaining proved (1P) reserves cannot reach the remaining 2P reserves until all the oil has been exploited! We can hardly rely on these "proved reserves", which are clearly at present largely exaggerated!

As for the reserves of shale (source-rock) oil, we have seen that there was no estimate possible without long-term exploitation. The amazing statements made on this subject are therefore not based on anything solid. The predictions of the EIA, which have run throughout the world, have been made from extrapolations to the rest of the world by simple rules of three of the example of American producer formations!

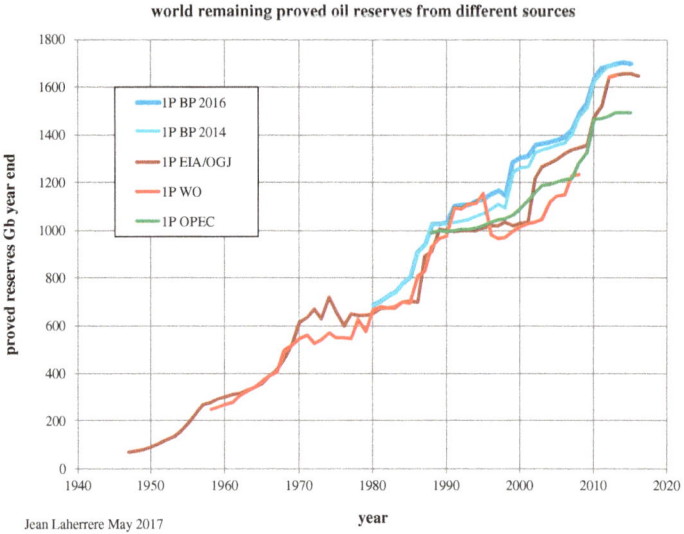

world remaining proved oil reserves from different sources

Jean Laherrere May 2017

Figure 38 *Various estimates of proved remaining oil reserves: Energy Information Administration and Oil and Gas Journal (EIA/OGJ), British Petroleum (BP), World Oil (WO) and OPEC.*

One could appreciate the value of this type of prediction for Poland, but even for the United States following the forecasts made for the formation of Monterey in California, which went from fabulous to virtually zero once we looked closer!

On the other hand, these publications of reserves do not specify at what price they will be extractable. However, there are currently very large differences in the break-even point of their production that is the price at which the exploitation becomes profitable. According to an estimate (Table 10) made at the end of 2015 for the anticipated oil production for 2020, this break-even ranged from about $ 10 to $ 90 per barrel depending on the oil category and its accessibility!

According to this table, at a price of $ 50 per barrel, most of these called oil reserves cannot be profitable, except for the onshore reserves in the Middle East.

All this shows that there is not much reliability in the "proved" reserves declared by the agencies and the companies and that it is safer to ignore them. Yet these are the only ones used by economists and the media.

3.1.2.2 The natural gases

The problems encountered in estimating gas reserves are exactly the same as in oil. However here there is no equivalent of heavy oils and bitumens[31].

---

[31] However, comparison with gas fields can be attempted, the exploitation of which requires unconventional techniques, such as those very rich in so-called acid gases (H_2S, CO_2), such as the giant Kashagan deposit in the north of the Caspian Sea, or those of methane dissolved in deep aquifers as in Louisiana in the United States.

| Table 10 | *Estimate of the break-even of oil production in 2020 according to its category and accessibility. Source Rystad-Energy. Morgan Stanley and U.S. Global Investors.* |

Category	Break-even fork, USD/barrel	Median break-even USD/barrel
Onshore, Middle-East	9-38	27
Offshore, continental shelfs	11-69	41
Extra-heavys	34-59	47
Onshore, Russia	27-71	50
Onshore, rest of the World	22-73	51
Deep sea	22-71	52
Ultradeep sea	38-65	56
Shale oil, North-America	52-75	65
Tar sands, Canada	49-86	70
Arctic	43-91	75

Figure 39 shows, for gas extracted from conventional deposits, the comparative evolution of the backdated 2P reserves and the proved published reserves.

The 2P reserves show a slight decline since 1990, followed by an increase following the very recent revaluation of the 2P reserves of the super-giant South Yoloten-Osman

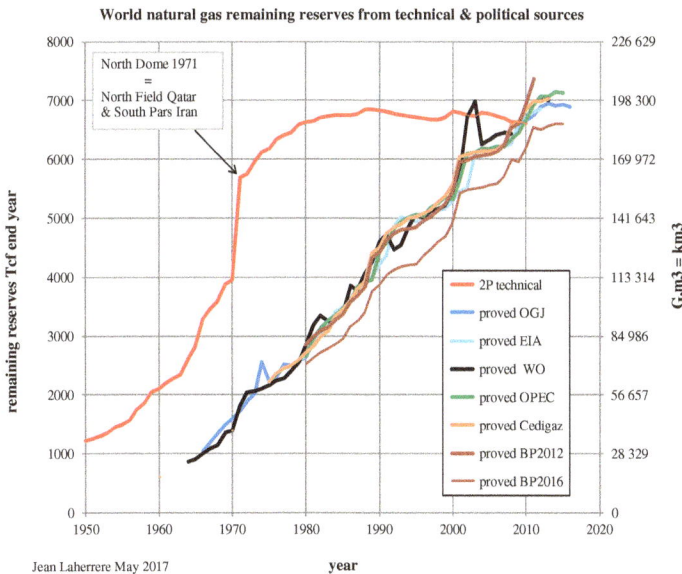

World natural gas remaining reserves from technical & political sources

Jean Laherrere May 2017

| Figure 39 | *Evolution of reserves 2P of conventional gas and comparison with "proved" reserves published by different agencies.* |

deposit in Turkmenistan. The same pattern is observed as for "proved" oil reserves: here too, the estimates of the remaining proved reserves have become approximately equal to that of the remaining 2P reserves in 2010, which is inherently impossible.

The 2P reserves of the largest known gas field, North Dome (2/3 Qatar North Field, 1/3 Iran South Pars), discovered in 1971, are estimated at 1500 Tcf, which is equivalent in energy content to 250 Gboe, 70% more than Ghawar.

The number of giants (> 3 Tcf) is 422 with 35 supergiants (less than for oil).

The gas Gini diagram (24,0750 fields with 9835 Tcf of 2P reserves) is very similar to that of conventional oil, with 60% for 162 fields (0.7%) and 90% for 2078 fields (8.4%). The Gini coefficient (0 perfectly unequal, 1 perfectly egalitarian) is even more unequal than that of conventional oil, of the order of 0.02! Here again, most of the reserves 2P are in a very small number of fields. However, these fields were found on average about ten years later than in the case of oil.

For unconventional natural gases (shale gas, tight gases, CBM, CMM), there is no more reliable estimate of reserves than for shale oil.

3.1.2.3 Coals

The two most authoritative sources of information are the World Energy Council (WEC) and the Bundesanstalt für Geowissenschaften und Rohstoffe (BGR) in Germany. In recent years, proved reserves assessments, which have been seen to be

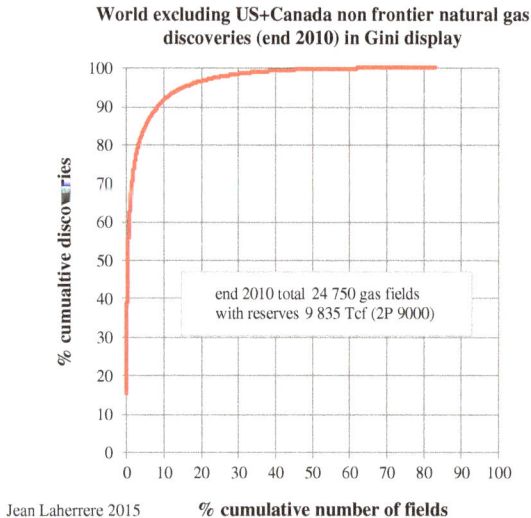

Figure 40 *Gini diagram of the distribution of conventional gas fields, excluding deposits in the United States and Canada non-frontier (i.e. including for the United States and Canada only deposits of Alaska and offshore).*

roughly equivalent to the oil industry's 1P reserves, vary from about 800 to 900 Gt for WEC and are around of 1000 Gt for the BGR. These reserves are made of many categories of coal of different calorific powers, and we must divide these values by about 2 to have the equivalent in Gtoe. The estimate of the BGR for 2013 (35) is 1052 Gt, of which 769 Gt of "hard coal", i.e. the addition of subbituminous, bituminous and anthracite, and 283 Gt of lignite.

Proved resources would be 3 to 4 times higher! But we have seen above that this notion, which does not correspond to specific projects, cannot be compared in reliability to the 2P reserves of the petroleum industry. As for resources in the broad sense, they would be about 10 times larger than the reserves, but these estimates have varied greatly over time.

This is therefore far from being here as reliable a work than for the oil and gas 2P reserves.

The relative persistence of proved reserves values cited over time (Figure 41) intrigues, while world production is currently about 8 Gt per year. It suggests that proved resources are transformed into proved reserves to some extent on administrative demand.

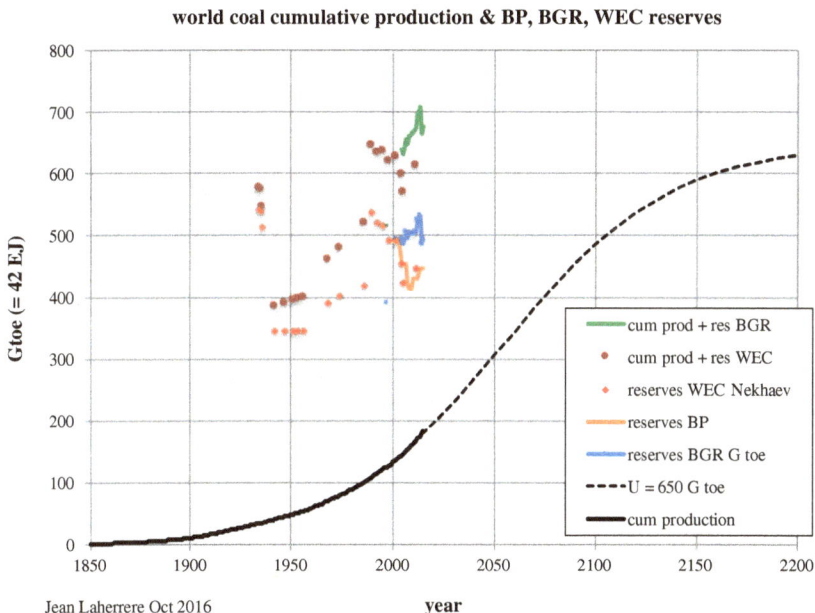

world coal cumulative production & BP, BGR, WEC reserves

Jean Laherrere Oct 2016

Figure 41   *Compilation of proved coal reserves, calculated in oil equivalent, by major sources of information, and comparison with cumulative production. The black dotted curve represents the approximate evolution of production, for ultimate reserves of 650 Gtoe (see Chapter 3.2.4), which corresponds to about 1300 Gt of coal. Courtesy Jean Laherrère.*

In any case, reserves depend in part on the price that society can pay for their extraction: the size of the coal reserves appears to be more sensitive to this price than conventional oil and gas reserves[32].

3.1.2.4 Bituminous (oil) shales and gas hydrates

We have seen for these domains the very speculative character of the resources displayed, and the absence of significant reserves. Technological advances and price increases, which would make exploitation profitable, may be able to change this one day, but this is not certain, as environmental concerns are becoming timelier. Again, the fear of global warming due to CO_2 emissions from the use of fossil fuels could lead to a moratorium on the implementation of these resources.

3.2 What is the future of global fossil fuel production?

Let us recall that predicting the evolution over time of global fossil fuel production is much more important in practice than knowing its ultimate reserves, even if ultimate reserves and production are linked in the long term. Indeed, as regards the progress of the world economy, it is the importance, year after year, of the energy flows which feed it, that it is necessary to estimate, in other words the possible flow of the carburettor of the car rather than the volume of its tank.

3.2.1 What do the Energy Agencies tell us?

To know and predict world productions, three sources are the most referred to: the Paris-based International Energy Agency (IEA www.iea.org), the US Energy Information Administration (EIA, www.eia.gov), part of the US Department of Energy (DoE), and the BP Statistical Review of World Energy (http://www.bp.com/statisticalreview), published by British Petroleum.

Another important source is the Organization of Petroleum Exporting Countries (OPEC).

Another example is the World Database (Data Publica, DB) of the Joint Organizations Data Initiative (JODI).

There is also the World Energy Council (WEC), a 90-nation expert forum based in London that holds the World Energy Conference every year.

---

[32] In economic terms, it will be said that the elasticity of supply of coal to a price increase is currently greater than in the case of conventional oil and gas. However, given the importance of CO_2 emissions and the air pollution caused by its use, the use of coal may be restricted internationally (see Chapter 5).

A big difficulty is that each of these sources has its own categories and calculation methods, and that, despite the confusion it creates, they still do not seem willing to synthesize their approaches.

The IEA represents rather the consumer's point of view, and its approach is mainly political-economic. Schematically, it addresses the problem of the future production of fossil fuels by demand, by making scenarios of it according to economic and political conditions, while implicitly assuming that supply will satisfy demand if prices are sufficiently remunerative. It makes annual compilations and forecasts in its World Energy Outlook (WEO). IEA publications are those most used by policy makers internationally.

The EIA has a more technical approach and rather represents the producers' point of view, that is, their production forecasts. It has its International Energy Outlook (IEO), and its Annual Energy Outlook (AEO) for US production.

The BP Statistical Review has the advantage of publishing historical series which are very useful for monitoring the evolution of productions over time.

From source to source, there is considerable dispersion (Figure 42) in published data for the production of "all liquids" petroleum (oil supply), which is the basic

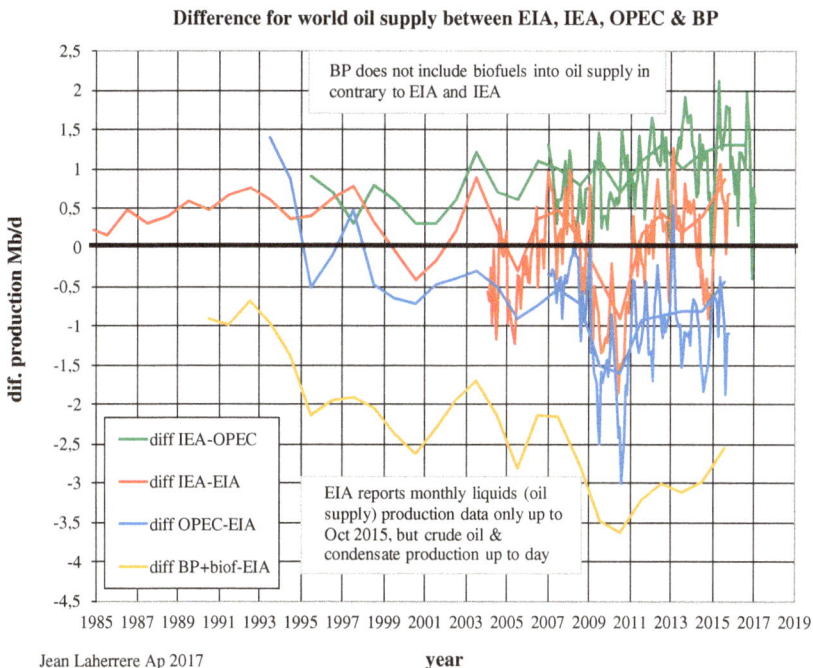

Difference for world oil supply between EIA, IEA, OPEC & BP

BP does not include biofuels into oil supply in contrary to EIA and IEA

diff IEA-OPEC
diff IEA-EIA
diff OPEC-EIA
diff BP+biof-EIA

EIA reports monthly liquids (oil supply) production data only up to Oct 2015, but crude oil & condensate production up to day

Jean Laherrere Ap 2017

Figure 42   *Comparison of estimates of global all-liquids oil production by different sources: EIA: US Energy Information Administration; IEA: International Energy Agency; OPEC: Organization of Petroleum Exporting Countries; BP: British Petroleum. BP does not take into account biofuel, so they have been added here to its data to make them comparable to that of other sources.*

datum taken into consideration by economists and the media. This is not surprising considering the difficulty of collating data, the differences in definition, inaccuracies, errors and omissions inherent in this type of approach, as well as the political pressures in a so sensitive subject. However, it is always good to recall the sources when we propose to comment on these data.

Differences from one source to another can be as high as 3 or 4 Mb/d, or about 3 to 4%, on the quantities of past and present production. This is about 3 times the French annual consumption. It should be noted, however, that BP, unlike the others, does not include biofuels in all liquid oil (oil supply). Those have been added in Figure 42 to BP data so as to make the comparisons from an agency to the other on the same basis.

3.2.1.1 The oils

We have seen in Chapter 1.1.3 the different categories of products that make up what is known as "all liquids" oil (also called oil supply or total oil), and their proportions in current world production. Let us note again that it contains categories that do not come from oil fields: - condensates and liquids from natural gas plants (NGPLs), sometimes grouped under the heading Natural Gas Liquids (NGLs), which are extracted from gas fields or gas associated to oil fields – a category which is not natural, synfuels, made of synthetic oil or gas – a category that is also not natural, refinery gains.

Agencies routinely aggregate condensates, i.e., liquid hydrocarbons from gas recovered at the wellhead, and conventional oil. Indeed they are often mixed by the producers on the field, and there is seldom separate accounting.

The international IEA combines conventional oil and condensates (C+C) under a conventional oil heading, and separately accounts for LTO and XH. And it often brings together all the synfuels under a single label.

In its most common publications, the US EIA adds LTO and XH to conventional + condensates to form a "crude oil" category. However, it is possible to find a separate accounting for the LTO and the XH and therefore to recalculate a category conventional + condensates. The EIA details the various synfuels.

It is therefore highly recommended, before commenting on the data published by the agencies, to carefully examine the categories they cover.

Between the two main agencies, the differences between the productions of all liquids oil are not negligible as we have seen, but above all, their medium-term forecasts are very dispersed and generally above production that is later observed (Figure 43). These agencies are showing an imperturbable optimism: according to them, the production of all liquids oil will be increasing constantly until 2030, 2040, and of course after.

The value of these forecasts can be also strongly questioned by analyzing one of the recent IEA reports, WEO 2012: Figure 44 shows the evolution of world oil

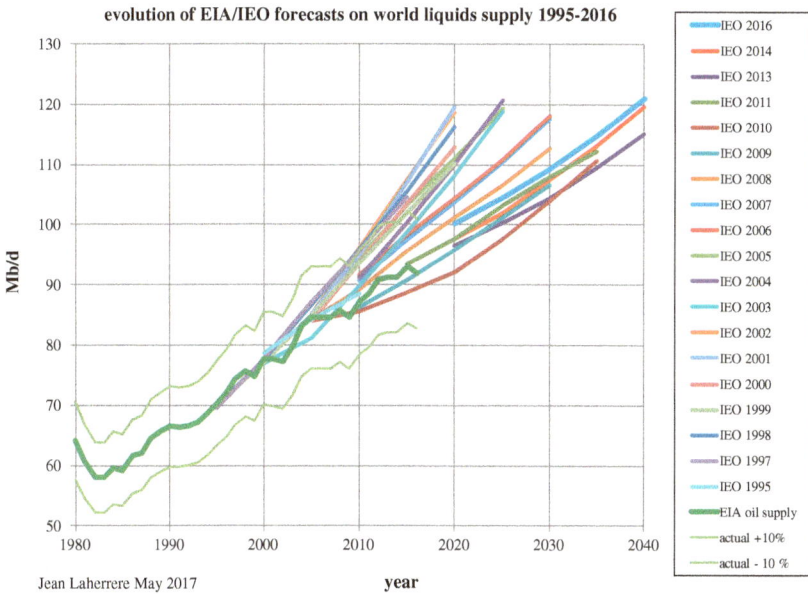

Figure 43 *Successive medium-term forecasts since 1995 of world liquid oil production by the EIA in its annual International Energy Outlook (IEO) and comparison with production to date. The wide dispersion of these forecasts in the medium term and their almost always optimistic nature are observed, despite revisions almost always downward from one year to the next.*

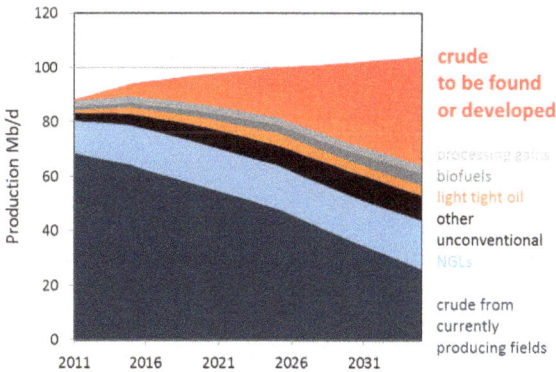

Figure 44 *Production projections for the various "all liquids" oil categories by 2035 according to the International Energy Agency (IEA) New Policies 2012 scenario in its WEO 2012. These forecasts are organized in such a way as to show the importance of the IEA betting in 2012 on future developments and discoveries of conventional oil fields. Crude: conventional oil. Processing gains: refinery gains. Courtesy P. Brocorens, University of Mons.*

production from 2011 to 2035 according to the scenario of this WEO called New Policies, which assumes an energy production constrained by environmental policy a little tighter than those of previous years. It should be noted that this scenario does not envisage any constraint on production that is geologic in nature. The WEO presents two other scenarios, one called Current Policies where there is no increase in environmental constraints, and the other called 450 Scenario, where on the contrary these constraints have become strong enough to impose a limitation at 450 ppmv CO_2 eq.[33] concentrations of greenhouse gases in the atmosphere at the end of the century (already about 485, of which 400 for CO_2 alone!).

The New Policies scenario seems to be the most representative of medium-term political and economic developments, and is in fact the basic scenario of the IEA. It should be noted, however, that the investment projects displayed by the energy companies would still correspond to the Current Policies!

The categories recalled above are recognized on these curves.

The world's all-liquids production would be in 2035 of 104 Mb/d, including about 7 Mb/d of synfuels and refinery gains, compared with 88 Mb/d per day in 2011, including about 3 Mb/d of synfuels and refinery gains.

It is particularly interesting to note that the expected growth in oil production is essentially based on that of conventional oil fields, with their share of total production however decreasing from currently 78% to 63%. But the share of conventional oil fields currently in operation (dark blue) would fall from 78% in 2011 to 25% in 2035!

According to this scheme, the share of the conventional oil that will have to be produced by other fields and technical progress (already discovered fields waiting to be put into exploitation, fields to be discovered, progress in the rate of recovery thanks to the technological progress) will compensate roughly the decreases of production of the fields currently in operation! Many observers believe that this is an impossible bet, since this implies a volume of conventional oil put into production by 2035 much too high, higher than the current cumulative production of the three main oil producers of the planet, Saudi Arabia, the United States and Russia! This contradicts the trend observed in the evolution of the world's 2P reserves in Figure 35.

The IEA forecasts, like those of the EIA, have in the past been almost always too optimistic! This seems to be the case once again. It should be noted that the WEO 2014 forecast is even more optimistic than the WEO 2012, 108 Mb/d in 2040. But the IEA has since put a little water in its wine, since WEO 2015 forecasts "only" 103.5 Mb/d on that date.

---

[33] The effect of non-CO_2 greenhouse gases (GHGs), mainly methane CH_4 and nitrous oxide N_2O, is added to that of CO_2 after having been quantified as equivalent CO_2 (CO_2 eq) in proportion to their effect in relation to that of CO_2. It should be noted, however, that the coefficients used in this quantification depend on the residence times of GHGs in the atmosphere, which are still sometimes very different estimates from one author to another, and also vary according to the term considered.

All this raises serious doubts about the ability of these agencies to predict the future of oil production, if only in the medium term.

3.2.1.2 The natural gases

The categories made by the agencies are here:

- Gas extracted from conventional deposits, associated and not associated with petroleum (categories 2 and 3 of Figure 28).
- Unconventional gas: shale gas, tight gas, CBM, CMM.
- Synthetic gas (syngas), from coking plants and blast furnaces, which represents little.

The associated gas produced by the exploitation of oil deposits is still frequently vented or flared due to insufficient transport or storage capacity or reinjected into the reservoir to maintain pressure or wait to higher gas prices. On unconventional deposits, re-injection is impossible in source-rocks or in a tight reservoir[34], and gas is often flared. This is the case, for example, with the gas associated with shale oil from the Bakken Formation in the United States, where about 1/3 of the associated gas is flared, compared with an average of 1% for the United States as a whole.

It is important to note that the IEA and the EIA accounts under the name of gas only for the production of natural gas having been processed in natural gas plants, i.e. what is called dry gas. It is not, therefore, the "gross gas", that is to say the totality of the gas at the outlet of the well, of which we have seen that a part was often flared or reinjected. This dry gas does not contain any more the natural gas liquids (condensates + NGPLs) contained in the processed production, which are, as we have seen, accounted for in "all liquids" oil. There is therefore a gap between gross gas and dry gas.

Figure 45 shows, for the United States, the range over time of the medium-term production forecasts for dry gas made by the EIA. As in the case of oil, the fragility of the medium-term forecasts of the agencies is shown year after year, and their wide dispersion.

Figure 46 shows the predictions of the IEA in its WEO 2015 for global gas demand by 2040 for the three scenarios already cited: the Current Policies Scenario, i.e. without an environmental policy, the New Policies Scenario, with more environmental constraints, and the 450 Scenario, where the obligation not to exceed 450 ppm of CO_2eq of greenhouse gases in the atmosphere at the end of the century is required. Except for the latter scenario, the demand for gas in dry gas (expressed here in billion cubic meters (bcm) under normal conditions, which corresponds to about 0.9 Mtoe) increases almost linearly until 2040.

---

[34] However, this re-injection could be considered in a conventional reservoir, if it exists nearby, or in a deep aquifer.

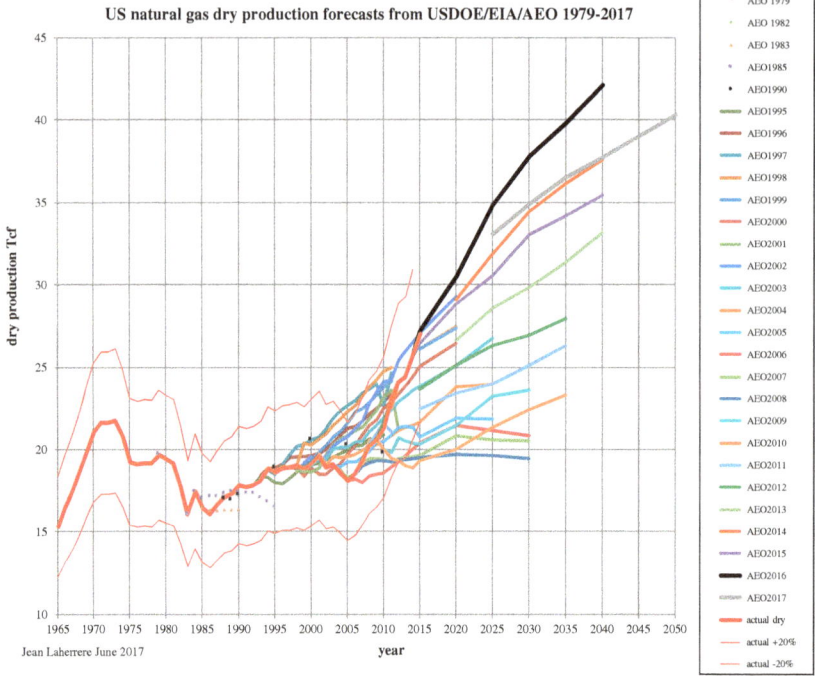

Figure 45 US dry gas production forecasts made by the EIA in its AEO from 1979 to 2015, and actual production curve. The rapid increase in production after 2005, not anticipated, corresponds to the "boom" of source-rock (shale) gases. The quantities are expressed in Tera (10^{12}) cubic feet (Tcf). 1 Tcf is about 24 Mtoe in energy content.

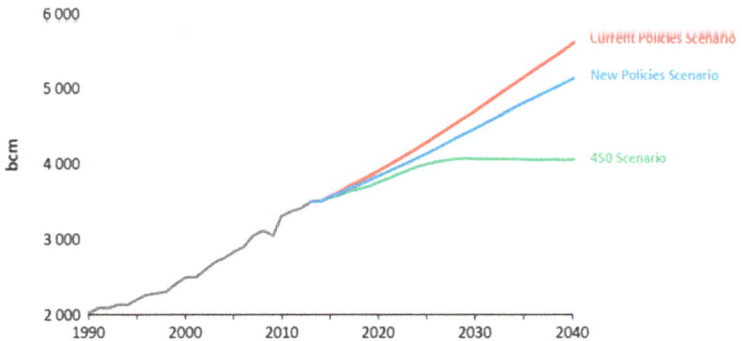

Figure 46 Forecasts in billion cubic meters (bcm) made by the IEA in its three scenarios of global gas demand by 2040 of its WEO 2015. Production is assumed to satisfy demand without problems. 1 bcm "is worth" in energy content about 0.9 Mtoe.

3.2.1.3 The coals

Table 11 shows the projections for coal production by 2040 under the three scenarios of the WEO 2015 of IEA. They are expressed here in terms of tonne-coal equivalent (tec), of which we have seen equivalent to 0.7 tons-oil equivalent (toe).

Table 11 *Forecasts of coal consumption and production by 2020 and 2040 by the IEA in the three scenarios of its WEO 2015. It should be noted that this is not a question of millions of tons of coal but of millions of tons coal equivalent (Mtce): these values are to be multiplied by 0.7 to obtain Mtep. There is also a relative consistency in the share of international trade (around 20% of production), OECD countries (mostly Australia, of course) being exporters to non-OECD. The lignite section also includes peat, which in fact represents little.*

		2000	2013	New Policies 2020	New Policies 2040	Current Policies 2020	Current Policies 2040	450 Scenario 2020	450 Scenario 2040
Demand	OECD	1 573	1 470	1 307	878	1 413	1 289	1 152	523
	Non-OECD	1 774	4 143	4 454	5 428	4 627	6 737	4 208	3 041
	World	3 347	5 613	5 762	6 306	6 040	8 026	5 360	3 565
	Steam coal	2 590	4 379	4 523	5 266	4 784	6 835	4 175	2 813
	Coking coal	452	940	929	785	941	851	903	601
	Lignite*	304	295	309	254	315	341	282	151
Production	OECD	1 380	1 361	1 255	1 042	1 391	1 505	1 134	627
	Non-OECD	1 875	4 362	4 507	5 263	4 648	6 521	4 226	2 938
Trade**	World	471	1 084	1 143	1 291	1 221	1 780	1 038	594
	Steam coal	310	814	847	984	913	1 447	759	373
	Coking coal	175	272	299	311	310	337	284	229
Share of world demand	Non-OECD	53%	74%	77%	86%	77%	84%	79%	85%
	Steam coal	77%	78%	79%	84%	79%	85%	78%	79%
	Trade	14%	19%	20%	20%	20%	22%	19%	17%

With respect to world production, only in scenario 450 it is expected to decline by 2040. As for the countries of the Organization for Economic Co-operation and Development (OECD), that is to say rich countries, the New Policies scenario also predicts a decline.

3.2.2 Methods other than that of agencies to predict the evolution of production: production peaks, ultimate reserves, creaming curves and Hubbert linearization

3.2.2.1 Peaks of production and ultimate reserves: historical examples: fossil fuels productions in France, the United Kingdom, Germany, the United States of America, and EU 28 + Norway

These analyzes of historical examples illustrate the fact that in a given country the possible production of a mineral substance is not infinite: it starts from zero and can only return to zero once the ultimate reserves have been completely consumed. Between the two, the production necessarily passes through a maximum, which is called peak production. This maximum may also be in the form of an irregular plateau that can last for many years.

The same can be said for world production, which is the envelope of productions country by country.

We are talking here about reserves, that is, quantities that can be exploited profitably under current economic conditions, rather than resources, a concept that does not take account of actual exploitability. The ultimate quantity of these reserves, called ultimate reserves or abbreviated to Ultimate, corresponds to the area under the production curve once it has been definitively stopped. We have seen that for conventional oil and gas, the best *a priori* estimate of this Ultimate was experimented to be what are called 2P reserves.

The Ultimate is very poorly known at the outset, and of course depends on economic and even political conditions (subsidies or tax exemption for example), not to mention environmental constraints[35] throughout the history of production, as well as progress in productivity through technology and organization, but it is becoming more precise as it is developed. We will see that there are methods to predict it in an approximate way if the exploitation is already well advanced.

France

French coal production began to grow quickly and steadily in the mid-19th century (Figure 47). After the vicissitudes due to the First World War, to the European economic crisis of 1933, a consequence of the Great Depression of 1929 in the United States, and to the Second World War, it went through a maximum in 1959. Its decline was immediately rapid. This is due to the fact that at that time production had become heavily subsidized compared to world market prices: this decrease was therefore accelerated in order to reduce the costs on the State budget. From 1974 onwards, after the first oil shock, the rate of decline was less rapid, with coal being more in demand. But the production eventually died out in April 2004 with the closing of the last active mines.

The history of gas and oil in France is long, starting with the heavy oil deposit of Pechelbronn and its satellites, whose surface indices were known for a very long time, and already used in the Middle Ages for lubrication and caulking, but also for their supposed medical properties. But there was no rational exploitation until 1740. It was in Pechelbronn that the first oil drilling for the exploration of the western world was carried out in the late 18th century and at the beginning of the 19th century, with borers which made it possible to reach a few tens of meters. The first to have

---

[35] In France, for example, all requests for the exploitation of coal in open pit are rejected for environmental reasons, however this is not the case in Germany, where gigantic lignite open pit exploitations are still operating.

France 1787-2015
Production of fossil fuels, Mtoe

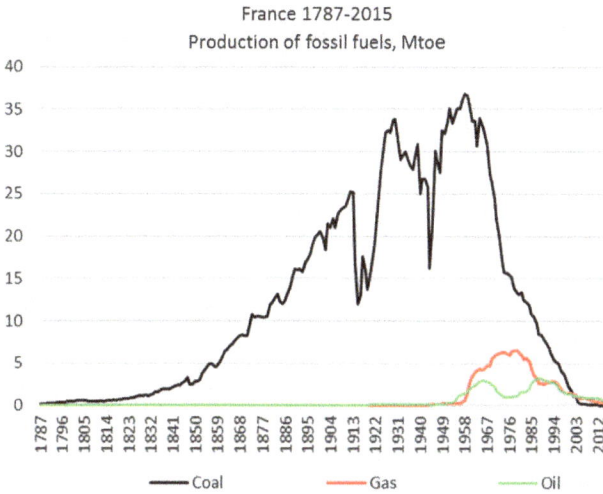

Figure 47 *Coal, natural gas and oil production in France from 1787 to 2015, reduced to the same unit of energy content (million tonnes of oil equivalent, Mtoe)). Data from O. Rech and BP.*

been officially recorded dates back to 1813, well before the drilling of Colonel Drake in Pennsylvania in 1859, the date chosen by history for the beginnings of the oil industry. France was even in 1850, adding to it the shale oil obtained by pyrolysis from bituminous shales, the Schistes d'Autun (whose first productions took place in 1824 at Igornay in Saône-et-Loire), the world's largest producer of oil "all liquids" of the time, until the explosive development of American production. These were however very small quantities, a few thousand tons per year (Figure 48).

Oil and gas production grew notably, but remained modest, after the Second World War. The oil was first extracted from the Basin of Aquitaine. This was then relayed by the Paris Basin, hence a production with two bumps. The peak dates from 1988 and the Ultimate is almost reached. The gas, which is mainly extracted from the Lacq field near Pau, peaked in 1978 and its Ultimate is almost reached.

New discoveries of gas and oil are not strictly impossible in France, particularly in the form of gas and oil produced from source-rocks (shale gas and shale oil). As we have seen, gas and oil have been trapped in or in contact with the sedimentary rocks which gave birth to them, in formations of very low permeability, but from which it is possible to extract some of them after hydraulic fracturing to increase this permeability. The corresponding reserves are supposed to be found mainly in the Paris Basin for oil and the South-East Basin for gas. Nothing is sure however, and the opposition of environmental associations is strong. On the other hand, the subsoil, in France, is the property of the State, which does not favor the involvement of landowners, who suffer from local nuisances without being associated, as in the United States, to profits from operations. Hydraulic fracturing in such plays was

**France oil production
from Pechelbronn & schistes d'Autun**

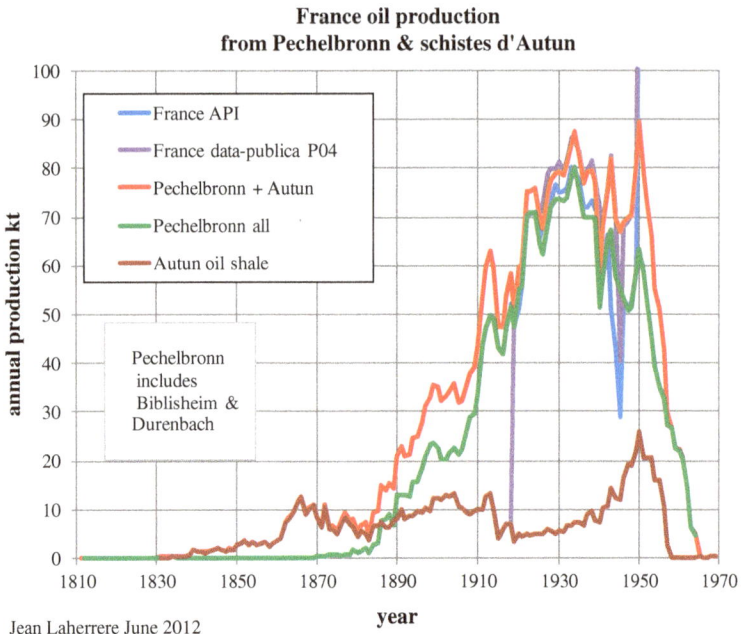

Jean Laherrere June 2012

| Figure 48 | *Chronicles of petroleum production in Northern Alsace (mainly Pechelbronn deposit) and the production of pyrolysis oil from the oil shale of Autun, according to various sources. These examples illustrate the evolution of production at the scale of the deposit, and the notion of ultimate reserves, which are the cumulated quantities extracted until abandonment of the deposit, i.e. the surface under the curves. Courtesy J. Laherrère.* |

banned in France by a law of 13 July 2013. In September 2017, prospection of oil and gas in France was banned by the government!

As for coal, there are still possibilities: in particular a rather large bituminous coal deposit (hard coal) (perhaps 250 Mt of reserves, equivalent to about 170 Mtoe) has not yet been exploited near Lucenay-lès-Aix in the South of the Nièvre department. However economic and environmental conditions do not seem to be favorable. And the mines that have been closed and are now flooded cannot be reopen as the cost would be far too high even for a market price that would be around 10 times the current prices.

The United Kingdom

The United Kingdom was the cradle of the Industrial Revolution at the end of the 18th century. This, before spreading to Europe and then to the rest of the world, developed there thanks to the coal of England and Wales, available in abundance.

Coal production grew rapidly and steadily to its peak, reached in 1913 (Figure 49). It then declined, but also registered severe drops of production due to the vicissitudes

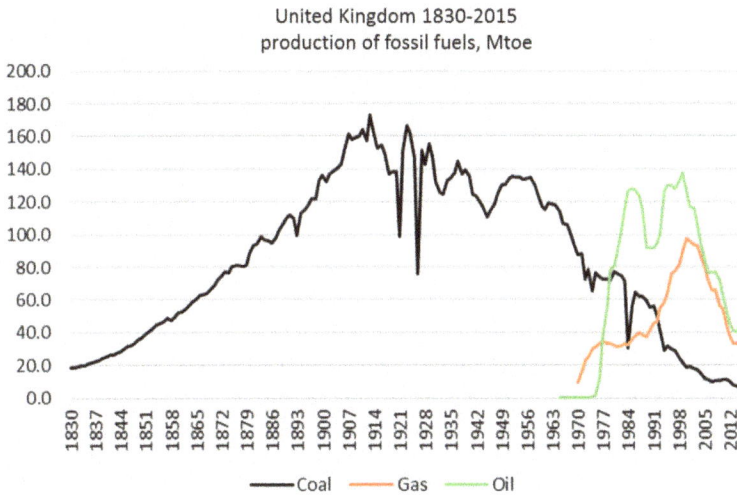

United Kingdom 1830-2015
production of fossil fuels, Mtoe

Figure 49   *Coal, gas and oil productions in the United Kingdom from 1830 to 2015, reduced to the same unit of energy content (million tonnes of oil equivalent (Mtoe)).*

of the history of this country: World War I, strike of the miners of 1921, Great General Strike of 1926, European Economic Crisis of 1933, World War II. After this, it was decided by the government, as in France, to reduce the high subsidies to its production. This led during the Thatcher era to the great strike of the miners of 1984. Production nevertheless continued to decline. The Ultimate is now almost reached. Note that in 1978, the World Energy Council foresaw for it 70 billion tons (Gt). In reality, it will be less than 30 Gt (see Figure 54).

The relay, very opportunely, was taken gradually from 1970 onwards by the production of natural gas and oil extracted from deposits discovered in the North Sea. These productions peaked in 1999 and have declined very rapidly since, despite high oil and gas prices from 2005 to 2014. A pause in the decline in oil production has however occurred since 2013, due to the large investments made in exploration-production during this period of very high oil prices, which then began to bear fruit.

In 1988, there was a sharp and temporary drop in production due to a very serious accident, the explosion of the Piper Alpha platform of the Piper deposit off Scotland, which disorganized a few years the oil production (and made 167 dead!). But this was not the main cause of the first bump in the oil production, which was mainly due to the exhaustion of a set of oil fields discovered at the start-up of exploration. Those were happily relayed by a set of fields discovered in a second step of exploration, which have now passed their peak of production.

In the United Kingdom, for all fossil fuel production, the peaks are largely passed, and the Ultimates seem close to being reached.

Particularly in the Bowland Basin in central England, exploitation of source-rock gases is envisaged, but for the time being it is impossible to predict the costs of any production.

Germany

Like France and the United Kingdom, Germany has coal deposits of good energy content (hard coal, about 0.7 toe per tonne on average) that can be mined by underground mines. It also has large brown coal (lignite) deposits, a coal of low energy content (about 0.25 toe per tonne in Germany) that can be exploited by open pit. This lignite can be produced at low cost and used to generate electricity near operating sites. The production of hard coal grew considerably and steadily from 1850 to 1914 (Figure 50). It then recorded the vicissitudes of German history: World War I, German economic crisis of 1924, European economic crisis of 1933, World War II. Its peak was reached in 1944. Then, as in France and the United Kingdom and for the same reasons, a decline was organized. Its Ultimate is almost reached: It is planned to close the last hard coal mines in 2018.

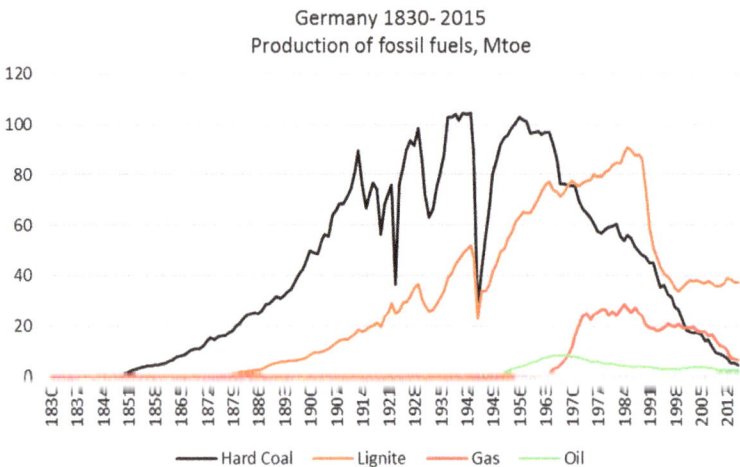

Germany 1830- 2015
Production of fossil fuels, Mtoe

Figure 50   Hard Coal, lignite, gas and oil production in Germany from 1830 to 2015, reduced to the same unit of energy content (million tonnes of oil equivalent (Mtoe)).

Lignite production started in 1870 and has experienced about the same hazards as coal, with less emphasis. However, it grew until 1989, the date of the German Reunification. It then fell very rapidly until 2001, due to the upgrading of the East German industry, which used a lot of lignite with low energy efficiency. It has risen again since 2001 and its Ultimate is far from being achieved. Germany is currently the world's largest lignite producer.

Natural gas and oil were first extracted from deposits mainly located in northern Germany. In a second stage, a small production of deposits located in the North Sea was added. The peak of oil took place in 1969, and that of gas in 1984. Even though production is not negligible at present, their Ultimate is now very close.

Germany is considering the production of gas and oil from source-rocks in Lower Saxony and North Rhine-Westphalia, but as in England and France, although it is not possible to know really the reserves. Such a production is strongly opposed by environmental associations.

The United States

Figure 51 shows the chronicles of fossil fuels production and consumption in the United States from 1949 to 2016. The data are from the US EIA and BP.

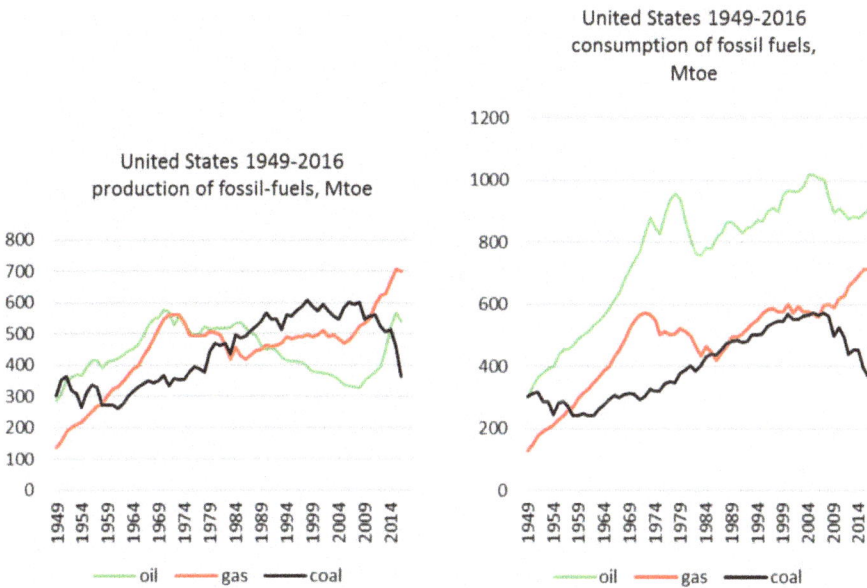

Figure 51   *Production (on the left) and consumption (on the right) of fossil fuels, in the United States from 1949 to 2016, in Mtoe. As concerns oil, it is the production of natural oil (C+C+LTO+XH+NGPL, see table 7), therefore excluding biofuels and other synfuels, while consumption includes those. However, this does not make much difference. Gas is so called dry gas, from which the NGPLs have been extracted. Sources: EIA and BP.*

The production of oil passed by a first peak in 1970. As we will see later on, this was predicted in 1956 by the American geoscientist Marion King Hubbert for the production of the so-called US Lower 48, i.e. the US States less Alaska and Hawaii. A secondary peak is observed in 1984, which corresponds to the peak of production of Alaska, which began to produce in 1967. Then is observed a steady decline, until 2008, when the production of shale oil begins to skyrocket till 2015. A small decline is observed in 2016, which corresponds to a drop in the production of shale oil. This decline has been confirmed in 2017. It is said to be due to the recent collapse in the market price of oil, this price being now insufficient to making profitable shale oil production. But also, as will be discussed later on, this could mean that production might begin to reach its geological limits.

Gas shows a quick increase up to 1973, then a decrease till 1986, a slow increase till 2000, then again a decrease till 2005. Then shale gas began to skyrocket, therefore some years before shale oil. Gas production is on a slight decline after 2015, for the same reasons than for shale oil.

As for coal a plateau is observed up to 1961, with a steady increase till 2005, then a strong decrease up to present day. This decrease is due to the replacement of coal in the electricity production, which is the main use of coal, by gas, and not to geological limits. Indeed, it coincides with the strong increase of gas production, which led to very low prices of gas in the United States, rendering gas-fired power stations more profitable than coal-fired. It is worth to note that a part of the excess production was then exported, in particular to European markets, where this coal on the contrary replaced gas, the price of which was much higher in Europe than in the US, in the production of electricity, thus leading to the closure or the mothballing of many gas fired power stations.

Consumption data show for oil two big "accidents". The first one is due to the oil "shocks" of 1973 and 1979 and the second one to the "subprimes" economic crisis from 2007 to now. The traces of the oil shocks are also visible on the chronicle of gas consumption, although less contrasted. However, gas is not affected by the subprimes crisis, and shows a quick and steady increase from 2005 to now.

As for coal consumption, there is a decrease till 1960, then a steady increase and a peak in 2006, the reasons of which have been explained above.

In 2016, the total consumption of fossil fuels represented 79% of the total primary energy consumption in the United States, and production 86% of the consumption. Therefore, the United States are not globally self-sufficient in fossil fuels. However, this is the case for coal and gas, while it is not for oil production, and never was since 1955. In 2016 its production represented about 60% of its consumption.

As compared to France, the United Kingdom and Germany, which are "has been" countries in what concerns fossil fuels, the United States are a "mature" country, which has not yet passed their peaks of fossil fuels production, but is probably close to reach them, at least for oil and gas, as described above. Its dependence of foreign countries is nil for gas and coal, but nevertheless very important for oil.

However, this rather favorable situation is quite entirely due to the strong development after 2005 of shale gas then shale oil productions. Otherwise, the productions of oil and gas would be largely on the decline, already from 1970 for oil! It therefore can be understood that the United States are anxious to see this "success story" going on for the years to come, but signs are now appearing of geological limits. It can also be understood that in European countries, at least oil companies and governments, if not environmental NGOs, are keen to see a similar success story starting in Europe. This will be discussed on our road to the conclusion of this book.

Europe of 28 + Norway

The Europe of 28 (EU 28) and Norway constitute a geographical group covering about 4.8 million km^2 for a population of about 515 million inhabitants. It is, for the moment, the second largest economic group in the world after the United States. Complete historical data for this group of countries as a whole are available from 1981 on, thanks to the BP statistical review of energy.

During this period, despite a strong decrease in coal consumption since 1986, and in oil and gas consumptions after 2005, the consumption has always been more important than production, in particular for oil.

All of the fossil fuels productions have now passed their peaks! In 1982 for coal, in 2000 for oil, and in 2004 for gas (Figure 52). While fossil fuels still account for 77%

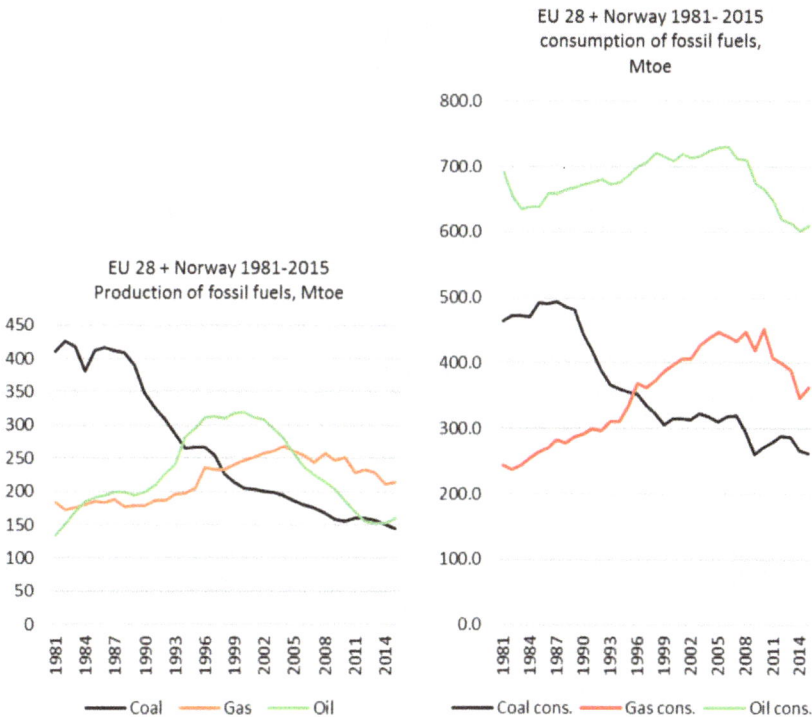

Figure 52   *EU 28 + Norway: on the left fossil fuels production, right fossil fuels consumption from 1981 to 2012, in Mtoe.*

of its primary energy consumption, i.e. a little less than for the United States, it is highly surprising that this does not appear to be of urgent concern to the governments of the member countries, given what this means for their energy dependence and the fragility of their supply.

Figure 53 shows the evolution of oil and gas production in the UK and Norway shares of the North Sea, which are by far the largest producing areas in Europe, from the first discoveries until 2015. For oil, the peak occurred in both countries in 1999–2000. It can be seen that the trend has not been reversed by exploration efforts, technological advances, significant price increases and tax exemptions from 2000 to 2013. However, the decline has been halted for oil and gas since 2013, but will this respite be sustainable after the recent collapse in oil prices?

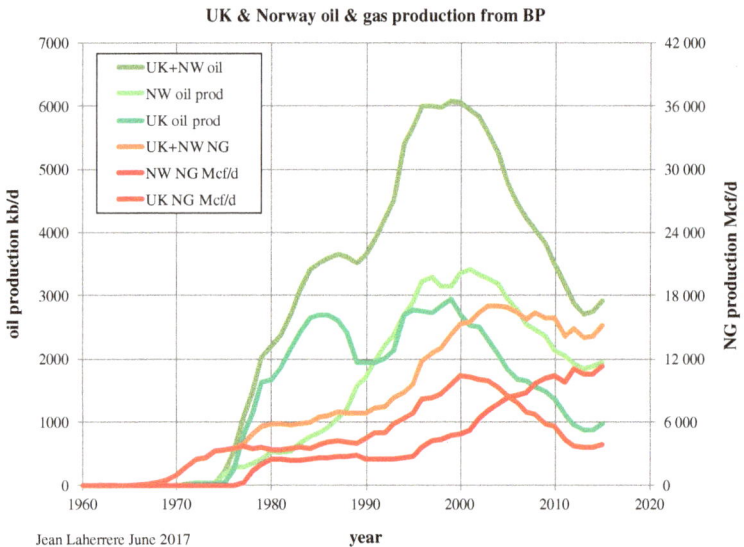

Figure 53. North Sea production history for the two major producers: United Kingdom and Norway, according to UK Department of Energy and Climate Change (DECC) and Norwegian Petroleum Directorate (NPD). It should be noted that these are conventional oil and gas, as these two countries only have deposits of this type in the North Sea.

The considerable temporary collapse in production on the UK side in 1988 was due as said above partly to the explosion of the Piper Alpha platform, which disorganized the operation for a few years, but more to the progressive exhaustion of a first set of exploration plays, fortunately soon relayed by a second one, which is itself now near to exhaustion. For gas, the peak was reached in 2000 on the UK side, and seems to be very close for Norway.

Regarding fossil fuels production, European countries as a whole have therefore now globally largely passed their peaks for oil and gas, and for coal also. Unless they are able to reduce quickly their consumption at the rate of their productions declines (but how?), they are condemned to a strong increase of their dependence of producing countries in the years to come. Exception are those of Norway for oil and gas, Germany for lignite, and Poland for coal, but not for very long.

3.2.2.2 Two methods for estimating the ultimate: the creaming curves and the Hubbert linearization

The creaming curves for petroliferous basins (oil and gas)

During the exploration of a petroliferous basin, one discovers in the course of time fields whose reserves are estimated. The construction of a so-called creaming curve consists in carrying over year after year, on the abscissa, the cumulated number of fields discovered, or rather the cumulative number of pure exploration wells, called wildcats (this graph which is a function of the real activity is much better than the field discoveries graph disrupted by the uncertainties of the exploration activity), and on the ordinate the cumulative of the corresponding reserves thus discovered. These reserves may be reassessed, but it is necessary to remain consistent that these revaluations are allocated to the revalued field at the very date of its discovery. This is called back dating, which could be also said of retroactive re-evaluation. On the other hand, as has been stressed, it is important not to use the so-called proved (also known as 1P, current or declared) reserves reported by energy companies or agencies that are tainted by many errors and voluntary or involuntary biases that render them unusable, inter alia because the correction of the previous estimate is attributed to the year of the new estimate, whereas the so-called 2P reserves should be attributed to the year of discovery. This requires a very patient collation work: it can be done from data published in the countries that give the reserves by field like the federal domain of the US in the Gulf of Mexico (BOEM) or Canada until 2009 with the CAPP, otherwise it is necessary to make scouting, form of industrial espionage by the companies selling the databases like IHS and Rystad. However, even these data have to be carefully analyzed, since they are produced from different classification systems (SEC, ABC, OPEC…) and are not homogeneous. They are also increasingly under the influence of producer countries or lobbies. Thus, for example, IHS became a few years ago IHS-CERA, owned by the Cambridge Energy Research Institute (CERA), based in Cambridge, United States, one of the lobbies of Western oil companies.

Over time, the cumulative number of discoveries increases, but their average size decreases because statistically we discover the largest deposits first. Cumulative reserves are therefore increasing less and less rapidly. This is due to the natural distribution of the sizes of deposits, according to which the bulk of the volumes are, as we have seen (Figures 36 and 37) in a very small number of very large deposits.

These graphs are modeled by curves whose asymptotes give an estimate of the Ultimate of the basin for the successive exploration cycles, these correspond to new approaches due to the progress of the techniques of exploration or exploitation. The same process can be used to estimate the Ultimate on a larger scale and for the world. Details of this method can be found in the works of *Laherrère (2011)*. Once the Ultimate has been estimated, a logistic function is set on the production history, which is bounded by the Ultimate. The part of the curve thus constructed between the last production data and the Ultimate is therefore an estimate of the average evolution over time of future production, excluding constraints other than geological constraints. The logistic functions were originally created by *Verhulst (1845)*, to predict population changes over time.

Figure 54 shows examples of creaming curves for the United States.

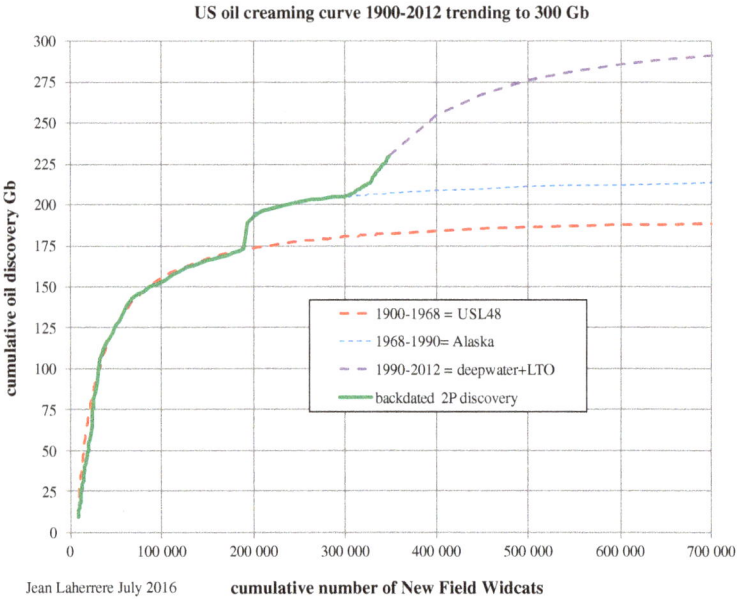

US oil creaming curve 1900-2012 trending to 300 Gb

y-axis: cumulative oil discovery Gb

Legend:
- 1900-1968 = USL48
- 1968-1990= Alaska
- 1990-2012 = deepwater+LTO
- backdated 2P discovery

Jean Laherrere July 2016   **cumulative number of New Field Widcats**

Figure 54   *Creaming curves for the United States: Creaming curve for the US 48 Lower states (i.e. excluding Alaska and Hawaii), to which is superposed the one for Alaska, which began production in 1968, then the one for oil extracted by deep sea and the Light Tight Oil (LTO). This is of course the 2P "backdated" reserves, which means that the revaluations of these reserves are always carried forward to the date of the discovery well of the deposit, not to that of the re-evaluation.*

Of the need for backdating

For those who have not really thought about the nature of what is called a reserve, it seems indifferent to postpone re-evaluations of reserves to the year of discovery. Is not this re-evaluation, whatever the date, a new quantity of oil which is discovered for mankind? In fact, this is not the case: Figure 34 shows how the representation of a deposit can evolve as a function of the number of wells drilled, and thus the calculation of the quantities in place, which we recall is a probabilistic calculation based on a geological description. So it is not the deposit that has changed, but the geostatistical representation that is made of it. The re-evaluation did not create new quantities, it simply modified the initial representation and therefore the probabilistic calculation of the quantities in place. The re-evaluations of the quantities in place are not always quantities added, they can be quantities subtracted. In fact, for many deposits, these assessments only begin to take a close look at physical reality after about ten years of exploitation.

However, the re-evaluation of 2P reserves must take account not only of the re-evaluations of the quantities in place, but also of the technological advances that have taken place in the meantime, which have increased the possibilities for recovery, as well as variations in the market price of oil. But these two factors play less and less in the case of conventional oil fields, as the example of the North Sea or the conventional oil fields of the US Lower 48 shows: With time the geological reality takes more and more over the economy and the technology.

As for the re-evaluation of the reserves 1P of an oil or gas field, this is the one which is published, which also does not bring by itself any new quantity: It is in reality only a game of writing that consists in passing the reserves from one category to another. And as this is tainted as we have seen with many biases and falsifications more or less voluntary, it is better to ignore it.

However the most important point is that if one does not assign re-evaluations to the year of discovery, it becomes impossible to observe the actual temporal evolution of the importance of discoveries. The curve thus constructed is not a curve of creaming, and it is impossible to attempt to predict from this curve the ultimate reserves.

A demonstrative text in this regard is that of *Laherrère, 2011*: "backdating is the key".

The Hubbert linearization

This method uses not historical discoveries, but production histories. It consists in plotting year after year the cumulative production on the x-axis, and on the y-axis the ratio as a percentage of the annual production to the cumulative production, and then in drawing a straight line representing at best the changes in production (Figure 55). It is a method popularized by the American geologist-geophysicist Marion King Hubbert (in fact by *K. Deffeyes*, who worked with him, in his 2001 book "Hubbert's peak"), famous for having predicted in 1956 that the production of conventional oil from the US Lower 48, i.e. excluding Alaska and Hawaii, would know, for ultimate reserves of 200 Gb, high estimate of a Delphi survey conducted at the time by the great American petroleum geologist Wallace Pratt, a peak in 1970, which actually occurred. At the time, there was little offshore drilling and neither extra-heavy oil nor shale oil was mined in the US. These were forecasts of the production of conventional oil deposits onshore and in shallow offshore. It should be noted that in the US Lower 48, as in Europe in the North Sea, which also produces only conventional oil, neither the new discoveries nor the technological progress nor

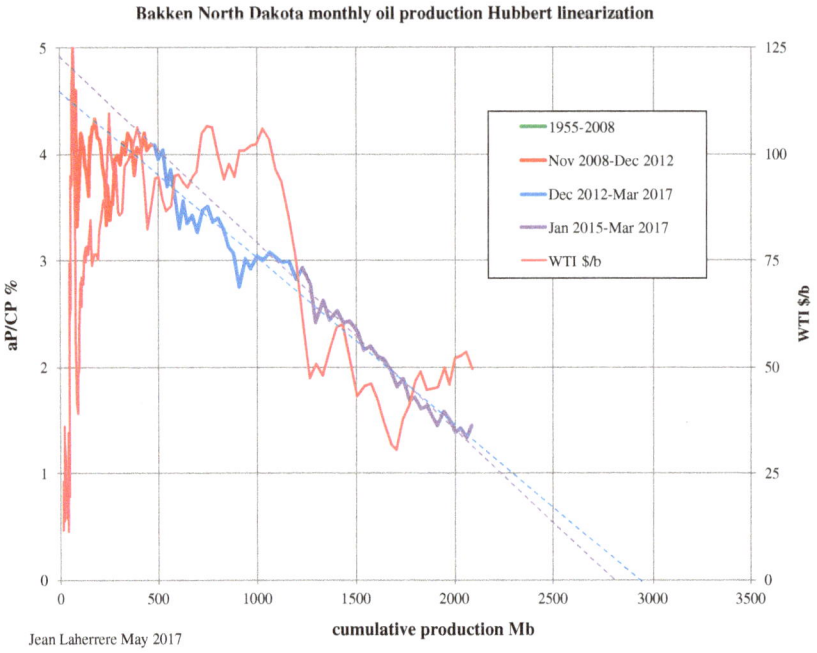

Bakken North Dakota monthly oil production Hubbert linearization

Jean Laherrere May 2017

cumulative production Mb

Figure 55 | *Hubbert's linearization of the production of shale oil from the Bakken shales of North Dakota, and price of the WTI crude during the same period. The fluctuations of prices has no real effect on the production.*

the considerable increases in market prices from 2003 to 2014, have really stopped the decline of this conventional oil.

This linearization of Hubbert allows, but less well than the previous one, to estimate the Ultimate. It is, however, useful to give a rough idea of the Ultimate in a basin that is already widely exploited, when the 2P backdated reserves estimates are not available, but also where classical methods of calculating reserves are not applicable: It is the case of shale oil and shale gas, but also of the ultimate reserves of extra-heavy oils and bitumens, where the methods of estimating reserves are, as has been said, problematic. It is the same with coals.

Figure 55 shows the example of the Bakken formation, the first in the United States for shale oil, for the part located in North Dakota. The extrapolation line of production histories intersects the abscissa axis a little below 3 Gb, giving the order of magnitude of the Ultimate of this formation. The EIA estimates it at 7 Gb!

Figure 56 shows the same exercise with coal production in the United Kingdom. The predictions here vary between about 22 and 28 Gt depending on the periods of exploitation, but it can be seen that as early as 1893, the prediction was close to the final reality. However, in 1978, the World Energy Council was still planning 70 Gt, about three times as much! It is interesting to note that the very sharp increase in coal prices that took place in the recent period in the wake of the oil price increase did little to boost production, just as increase in oil price has hardly changed the rate of decline in conventional oil production, neither for the US Lower 48 nor for the North Sea.

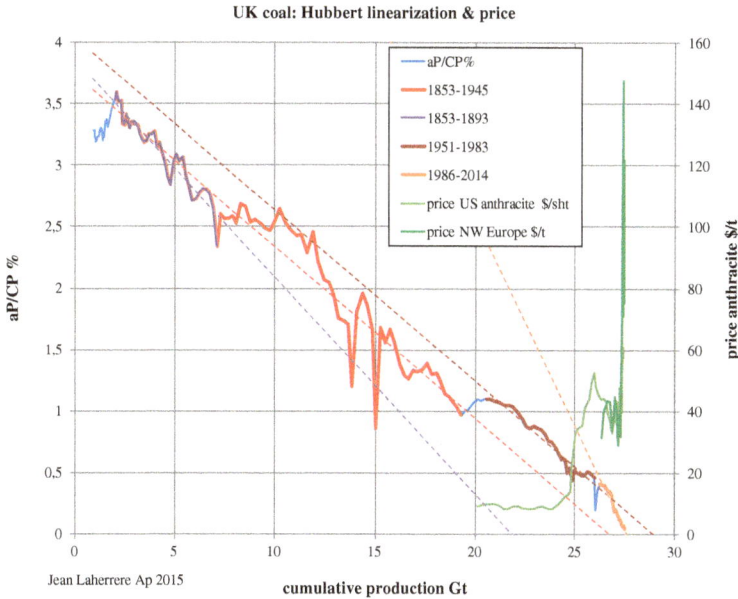

Figure 56 *Hubbert's linearization predictions of the ultimate coal production in the United Kingdom at different times of production, and changes in coal prices over the recent period. It can be seen that the very large price increases did not affect the observed decline significantly.*

However, it is necessary to be careful with the Hubbert linearization for which the conditions for a correct application are not always present, especially when the production histories are too short and then the cumulative productions are still only too weak a proportion of the ultimate. This can be seen clearly in Figure 55, where an estimate made when this proportion was less than 1/3 would have been hazardous. During operation, this method becomes increasingly reliable.

Figure 57 is a synthesis of predictions, estimated in Gtoe, of the future global production of all liquids petroleum, dry gas and coal made by J. Laherrère using these methods.

The ultimate reserves would be of the order of 390, 300 and 650 Gtoe respectively. The peak of all liquids oil would occur around 2020 at about 4.5 Gtoe (95 Mb/d). For gas, this would be about 3.6 Gtoe by 2030, and for coal, a peak of about 4 Gtoe already passed in 2013. The peak of total fossil fuel production would be around 2025 for about 11,4 Gtoe (475 EJ) of energy content. Within a decade, the total

137

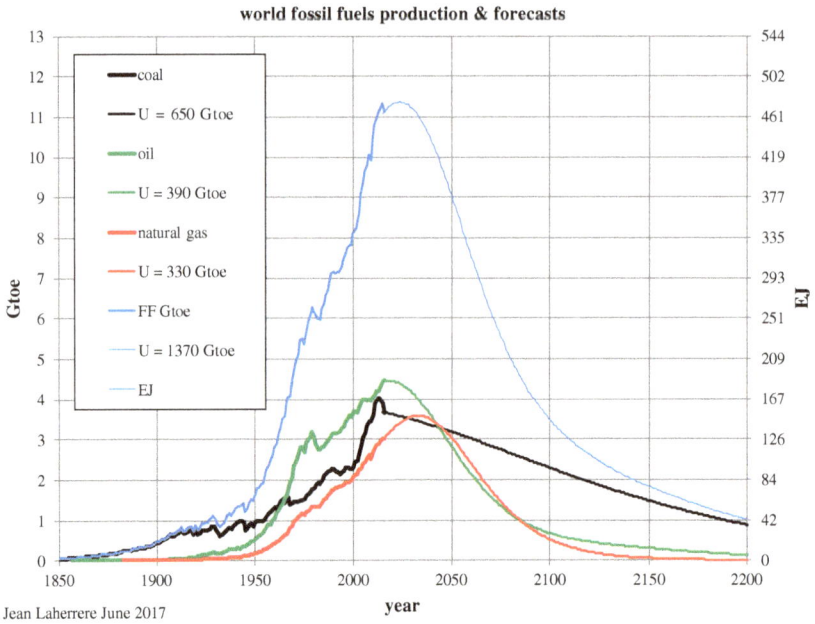

Figure 57 *Total petroleum (all liquids), gas, and coal production, and the total of three, in Gtoe and Exajoules (EJ), and production forecasts up to 2200 according to J. Laherrère.*

amount of energy contained in the production of fossil fuels would thus begin to decline!

The case for coal is interesting: for decades now it as been claimed that coal would fuel an increasing world population unceasingly healthier for at least the two next centuries. Here we see how dream is far from reality. Despite the shape of the future production curve is still not very well defined, it is likely that the global peak coal has been passed recently, mostly due the sharp decline of the Chinese production since 2013.

It has to be understood that these predictions are only by taking into account geological constraints (underground) and in the absence of constraints above ground other than those due to the current global economic and political fluctuations, known as Business as Usual (BAU).

The curves in Figure 57 are the global envelope of oil (all liquids), gas and coal production. These, in particular those of oil, include, as we have seen, several categories. These categories have been studied separately beforehand, because each of them has a dynamic of its own production evolution, depending on its origin, its physical characteristics and its methods of production. We have already mentioned that, for example, the speed of production of extra-heavy oil reserves (Venezuela's extra-heavy oils and bitumen from Canada) could only be lower than that of conventional petroleum reserves, their exploitation being more complex and expensive. For this reason, it was necessary to make separate accounts. The economy also plays a bigger role here than it does for conventional oil, because the exploitation, being expensive, is more sensitive to market price of oil than for conventional oil. The recent collapse of the market price of oil caused several Canadian bitumen operations to be abandoned.

For all liquids oil, the predictions by category are shown in Figure 59.

The Ultimates of fossil fuels would total about 1370 Gtoe in round count (Figure 57 and Table 12). Approximately one-third would have already been consumed, almost half for oil, and almost 30% for gas and coal (Table 12).

Table 12   *Ultimate 2P reserves (Ultimates) and 2P reserves remaining to be exploited, according to J. Laherrère 2015. Note that these reserves are evaluated in energy content (Gtoe) and not in mass or volume.*

Fuel, Gtoe	Oil (all liquids)	Gas	Coal	Total
2P reserves (Ultimates)	390	330	650	1370
Remaining 2P reserves, at the end of 2014	210	237	473	920
Cumulative production at the end of 2014	180	93	177	450

The comparison with the IEA 2012 WEO 2012 New Policies (NP) scenario, which estimates 104 Mb/d in 2035 (Figure 44), or about 4.9 Gtoe, shows a very large deviation from the Figure 57 predictions for all liquids petroleum production at this date. The WEO NP 2015, is very little less optimistic than the WEO 2012, since it predicts "only" 103.5 Mb/d in 2040. Then, J. Laherrère expects about 3.2 Gtoe, or about 67 Mb/d in 2040. This is a gap that is almost 4 times the current output of Saudi Arabia! On the other hand, this WEO do not foresee a peak before 2040, while Laherrère expects a peak in 2020!

For gas, the WEO 2015 NP and J. Laherrère forecasts are close for 2030, but then diverge: the WEO 2015 NP forecasts 182 Tcf (equivalent to about 83 Mb/d of oil) in 2040, and does not foresee a peak. J. Laherrère only expects about 150 Tcf (the equivalent of 68 Mb/d of oil) in 2040, and a peak in 2030. There is also a considerable gap here.

As for coal, this is where the predictions should be the most uncertain: Laherrère predicts for 2040 the oil equivalent of about 68 Mb/d while the scenario WEO NP 2015 predicts about 90 Mb/d of oil equivalent.

Agencies' expectations seem optimistic. The predictions of J. Laherrère, on the contrary, are they too pessimistic?

Recently, a Delphi survey conducted among the members of the French section of the ASPO shows that, on average, its members are more optimistic than J. Laherrère not on the date of the oil peak, but on the rapidity of the decline, that they see slower than does Laherrère.

Other forecasts are, for example, those of Yves Mathieu 2011, of the Institut français du pétrole et des énergies nouvelles (IFPEN), which envisages a peak or plateau of all liquids oil between 2020 and 2030 at heights of 95 to 100 Mb/d depending on the importance of technological efforts. This is not very different from Laherrère's. The Cambridge Energy Research Associates (CERA) estimates in the United States are on the contrary even more optimistic than those of the Agencies.

The big Western oil companies do not communicate much on this subject. Some, BP, Exxon, have also made estimates, which are generally more optimistic than those of ASPO, but the French Company Total predicted some years ago (*J.M. Masset, 2009*) a peak of conventional plus non-conventional oil a little before 2020, for about 95 Mb/d, and *P.R. Bauquis, 2014*, of the same company predicted a peak or plateau of all liquids oils between 2020 and 2030 to about 100 Mb/d, which is above but still close to the average forecast of ASPO members.

But the methods and models used by the companies and even by CERA, which is very close to them, have not been sufficiently explained by these sources to be discussed here in detail.

3.2.3   *Global production forecasts for this century, a discussion*

In brief, two categories of methods are therefore well documented in practice to try to predict the evolution of global fossil fuel production:

- That of energy agencies, which is essentially a modeling of demand according to economic and political criteria.
- That of ASPO geologists and economists, which is essentially geological modeling based on an in depth study of the histories of discovery and production.

Which of these methods will be the most predictive?

What about forecasts of agencies and companies?

With regard to oil, that of fossil fuels, which should be the first to reach its limits according to all the forecasters, the reserves they publish are declared reserves, in principle reserves 1P, which we have seen very unreliable and very optimistic. Agencies and companies should be able, if they really wish to do so, to publish 2P backdated reserves which are, as we have seen, much closer to the ultimate reserves, but they do not.

For our concern here, that is, the future of all liquid oil production, the predictions of the agencies and of most companies have been almost always too optimistic in recent years: The reasons for this are that they use models that set no limits on the possibilities that nature offers to man with regard to fossil carbon energy production flows, provided that he has the financial means, which means a subordination of the primary energy production flows to the cash flows, and thus implicitly ignoring the geological constraints! Yet the European historical examples that have been presented are a clear contradiction of this view: despite all the efforts of exploration, technological improvements and very high prices in recent times, or even of tax exemption as for example in the British North Sea, in none of these cases the trend has been really again on the rise once the peaks have passed. The same was true for US Lower 48 conventional oil production, and for a large number of producing countries.

It may also be thought that the optimism of agencies and companies also comes from the fact that they can hardly present a picture that is not encouraging to their sponsors, to their shareholders and to opinion.

The same causes produce the same effects, it is hard to believe that the predictions of oil companies and agencies will not continue to be too optimistic.

The same applies to natural gas and coal.

What about geological models?

These models are based on reserves estimates that are much more credible than those used by agencies, but a weak point is the precision of the estimate of ultimate reserves (Ultimate, U).

We compare below the analyzes of future productions for the various components of all liquid petroleum.

The oils

Figure 30 and Table 7 show the different constituents of all liquids petroleum and the relative importance in Mb/d of their production in 2014.

Let us note again that there is a category, the synfuels, which has nothing to do with natural oil, and another, refinery gains, which is a consequence of volume accounting and does not make sense from an energy point of view.

The dynamics of production we have said are different from one category to another and must be taken into account in the analysis of production possibilities in the future.

Conventional Oil and Condensate (C + C)

This is oil extracted from the conventional types 1 and 2 deposits in Figure 28, and condensates that are liquid hydrocarbons recovered at the wellhead from gas: gas associated to conventional oil and gas-caps, or gas from conventional gas fields containing liquid hydrocarbons (Part 1, Chapter 1). The dynamics of condensate production are thus linked to that of gas production, not to the production of conventional oil.

Agencies gather conventional oil and condensates in the same category, as they are mixed by producers before marketing. However, if the IEA presents separate C + C accounts, EIA associates it in its current publications with light tight oil (LTO) and extra-heavys (XH), i.e. extra-heavy oil + bitumen, in a category crude oil or petroleum. However, the EIA also publishes special statistics for LTO and XH.

The proportion of condensates in the C + C set is therefore not indicated, although the IEA sometimes attempts to do so, but with little precision. It has increased over time because, until around 1970, gas was often considered a fatal product because of the very high cost of the infrastructure needed to transport it, resulting in many of the discovered deposits being abandoned. At present, it could be in the range of 7 to 8%.

The quality of the results obtained by the geological models is certainly as we have seen sensitive to the quality of the estimates of ultimate reserves (ultimate, U). However, with regard to this C + C category, the production predictions thus produced have given results far closer to subsequent observations than those of the agencies, CERA and the majority of the oil companies, the best known being the prediction for 1970 of the US 48 conventional oil production peak, made in 1956 by Marion King Hubbert for an ultimate of 200 Gb.

Globally, Campbell and Laherrère 1998 predicted that the "cheap oil" peak, roughly the C category, would be reached at the world scale within 10 years. According to the IEA, this peak occurred at the end of 2005 and the decrease in production since that date reached about 4% in 2015 (Figure 58). There is therefore a fairly good agreement of the observations with the predictions of Campbell and Laherrère.

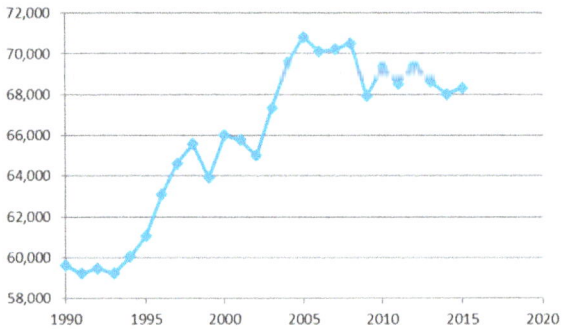

Figure 58 Evolution of world conventional oil production (category C) from 1990 to 2015, in millions of barrels per day, according to the International Energy Agency (IEA). In fact, the decline is probably a little more pronounced, because some of the condensates (from the gas, let us recall) has been counted here with the oil in the accounts of the producers. Courtesy P. Brocorens.

The peak production of conventional + condensates from the North Sea was also forecast for the exact year by ASPO and the forecasts were also good in countries that have now passed their peak, when it was possible to properly collate good data (Saudi Arabia, Russia, and China are examples of countries where it is difficult)

Finally, there is statistical evidence, year after year, of the decline in the volumes of conventional oil discovered. They are currently less than half the consumption. Despite the great advances in exploration techniques no giant field (i.e. 2P reserves greater than 500 Mb) was discovered in 2013, 2014 and 2015!

The maintenance of C + C production at 2035 and even more at 2040 at the current level, as the IEA claims (Figure 44), seems therefore very unlikely, and the page seems to have been turned in 2005-2006, in accordance with the prediction of geologists.

This is about 75% of the current production of all liquids oil!

Light Tight Oil (shale oil and tight oil)

Because of the current inability here to evaluate reserves 2P on geological and geophysical bases, the Hubbert linearization of production histories must be used (Figure 55). And these are still a little short in the United States, the main producer country, to be very reliable.

Production in 2014, about 4.4 Mb/d, is modest, about 5% of world production of all liquid petroleum (Tables 7 and 13), and is essentially limited to that of the United States. It is currently declining, as current oil prices have become too low for part of their production to be profitable. It may be temporary, as prices of oil are now recovering, giving a new impetus to production. However, some analysts were already predicting before the price drop the peak of the shale oil in the United States before 2020, including oil geologists who analyzed in great detail the characteristics and productions of all North American formations (*Hughes, 2014*, and the Enno Peters website: https://shaleprofile.com/). It should also be noted that this production has led to huge financial losses for many operators. These have been kept afloat by bank loans and tax deferrals or credits, that is, a financial bubble and indirect subsidies that may not last indefinitely.

A peak or a plateau at about 5 Mb/d around 2020 in the United States seems likely (*Chavanne, 2006*).

Can we consider substantial developments at the international level? A recent review of the question can be found in *Charlez (2016)*. For the time being, only Argentina with the operation of the Vaca Muerta shales in the Neuquen Basin (see Chapter 3.1.4.2 of the first part) has begun to record some successes, and is announced to progress quickly in the near future. There are still a few things, about 58,000 b/d at the end of 2016. However this production might be doubled end of 2018 according to *Wood and Mackenzie (2017)*.

In Russia the formation of Bazhenov, source-rock of the oil fields of Western Siberia, is predicted to a great future. China and Australia made some attempts of production, and there could be a large potential in source-rocks of Saudi Arabia, in particular the source-rock of Ghawar, the largest oil field in the world. The future will tell if this development will take place. It should be noted that the obstacles to be overcome are just as politico-economic as they are geological. It should also be noted that Russia and the countries of the Middle East still have large reserves of conventional oil at low production costs and have no immediate interest in embarking on LTO production, which is more costly to produce.

Extra-heavy oil (Extra-Heavy, XH)

These are mainly extra-heavy oils from the Orinoco tar belt in Venezuela and bitumen from Athabasca in Canada. It is impossible to construct creaming curves here, since they are singularities, and the notion of reserves is unclear, since they depend too much on exploitation techniques. Hubbert's linearization must therefore be used. It is relatively reliable because the production histories are already long. Possible production is fairly well understood: we have seen that it was handicapped by extraction processes with limited productivity and strong environmental constraints.

J. Laherrère's predictions are here of a slow increase in production to about 16 to 17 Mb/d by 2060, followed by a slow decline until the end of the century (Figure 59).

Liquids from natural gas processing plants (Natural Gas Plant Liquids, NGPL)

These are the "liquids" recovered from the gas in the gas plants plus LPG, i.e. after the condensates have been recovered at the wellhead. They are mainly LPG, i.e. propane and butanes liquefied by refrigeration.

As their name suggests these liquids are associated with natural gas, and their production dynamics follows that of conventional and unconventional gas. As far as the latter is concerned, it can only be NGPLs extracted from shale gas, the other unconventional gases containing little hydrocarbons other than methane.

According to J. Laherrère (Figure 59), an increase to about 12 Mb/d is expected around 2030, followed by a decline. A very large expansion of the world's shale gas production would delay this timeframe, but is it possible by then?

Synfuels and refinery gains

These categories are not natural oils, and their production cannot therefore be predicted by geological methods.

As for synfuels, they are mainly biofuels, whose production is of the order of a little more than 2 Mb/d currently (Table 7). This production is limited by the very low yield of photosynthesis (on average, less than 1% of the solar energy is converted by the plants into chemical energy on the surface they occupy) and thus by the surfaces

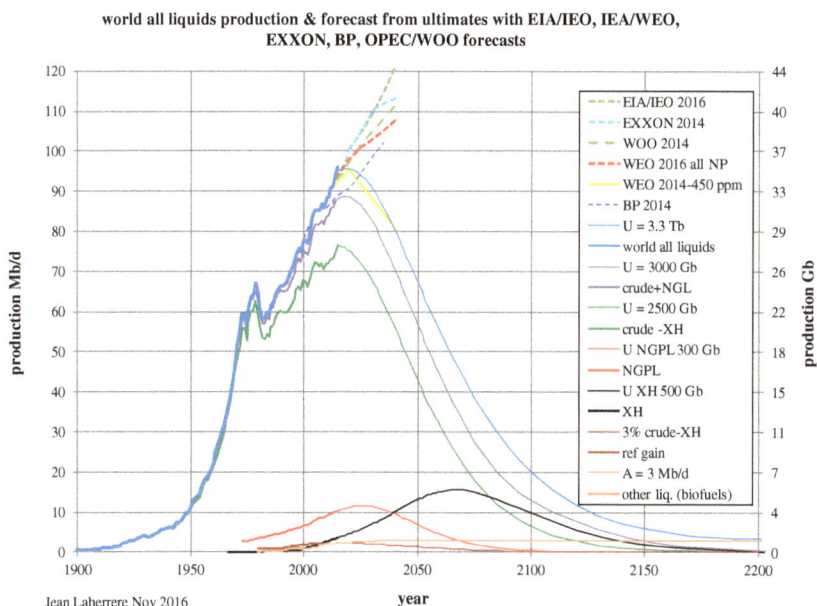

world all liquids production & forecast from ultimates with EIA/IEO, IEA/WEO, EXXON, BP, OPEC/WOO forecasts

Jean Laherrere Nov 2016

Figure 59 *Forecast 2016 of J. Laherrère. This Figure uses the EIA codes and requires a lot of gymnastics to decrypt it: production chronicles and forecasts of Laherrère for: crude oil -XH (= C + C + LTO), in dark green; crude oil (= C + C + LTO + XH) + NGPLs, in purple; all liquids oil, in blue and comparison for all liquids oil with the recent estimates of Exxon, BP, OPEC (WOO: World Oil Outlook), EIA (IEO: International Energy Outlook) and IEA (WEO: World Energy Outlook). NP: New Policies. U: ultimate reserves; XH: Extra-heavy (bitumen and extra-heavy oils); NGPLs; Biofuels; Refinery gains. The WEO 2016 all NP is the IEA's all liquids oil forecast until 2040 for the New Policies scenario. It can be noted that this forecast is for 2040 higher than that of J. Laherrère by about 25 Mb/d, i.e. more than twice the present production of Saudi Arabia.*

available for crops: It takes huge areas to get significant quantities. This poses major problems of land use, competition with food-based agriculture, and the environment. It can be expected, therefore, that production will hardly exceed the current level in this century, unless there is a revolution in the techniques of transforming biomass into fuels by BTL techniques, making it possible to put much more fully to this end biomass, and in particular its lignocellulosic component (wood, leaves, roots…), much more abundant than the agricultural products used until now. But this type of production does not currently go beyond the pilot stage. It should also be noted that the majority of biofuels are ethanol, whose energy per barrel is only two-thirds of that of conventional oil.

Those of GTL and CTL are in total less than one Mb/d at the present time. They correspond to a transfer, with a considerable loss of primary energy (and also with a high production of CO_2 and air pollutants!), from the gas and coal categories to

145

the oil category, and thus do not increase the total ultimate reserves of fossil fuels, but make them consume more quickly. In this regard, it should be noted that in the global energy balance, gas and coal used to make GTL and CTL continue to be accounted for by the agencies in the gas and coal production sectors, while GTL and CTL are counted in the oil production category. There is thus double accounting and illusion of an increase in the production of primary energy. As for their possible production, we cannot say much at the present time, because even when oil prices were very high, their production did not increase significantly. They are "productions of the future", this future being indefinite.

The refinery gains are variable depending on the nature of the oil entering the refinery and the relative importance of the oil cuts produced. They can be considered roughly as proportional to the quantities produced of C + C + LTO, and they will therefore follow their evolutions. In Figure 59 they are estimated at 3% of these quantities, and should pass through a maximum of 2.5 to 3 Mb/d in 2020.

We have seen that this was in fact a consequence of volume accounting, and indeed a loss of energy. This oddity leads to absurdities, among others the classification by agencies as oil-producing countries of countries that only refine it, such as the Netherlands Antilles, which refine oil from Venezuela.

All liquids Oil

It is therefore the sum of the previous categories. Figure 59 summarizes the forecast of its evolution by J. Laherrère at the end of 2016, as well as the evolution of its main components, and compares it to that of the main agencies, of OPEC, of Exxon and of BP, which makes it possible to visualize the considerable discrepancies of his forecasts with their forecasts, and this in the short-term.

Let us note in this Figure that the scenario WEO 2014-450 ppm (in yellow) is a scenario 2014 of the IEA where all liquid oil production would be constrained by a very rigorous international policy of reduction of CO_2 emissions. It is curious that it is in fact very close to the scenario of Jean Laherrère, which is a business as usual (BAU) scenario, where no constraints other than geological constraints are on CO_2 emissions. We will discuss this in Chapter 5.

What does the current evolution of the production of all liquids petroleum tell us?

Outside the United States, the estimated crude oil + condensate production according to the EIA (i.e. recall of C+C + LTO + XH), evaluated in barrels, is still increasing at the moment. However, there is a decline over time in growth rates, which suggests the arrival of a peak in this category (Figure 60). The growth of all liquids is mainly due to the growth of NGPLs produced from gas.

It can be seen in this Figure that the United States has experienced a very rapid increase in their all liquids production since 2010 thanks to their production of

World excluding US oil production from EIA

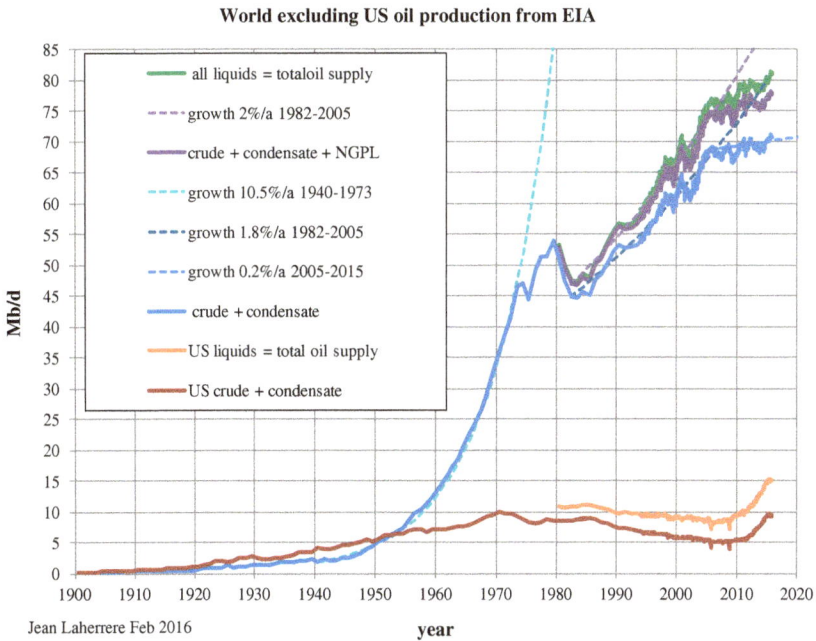

Jean Laherrere Feb 2016

Figure 60 — *Evolution according to the codes of US EIA of world production outside the United States of crude oil plus condensates (= C+C+LTO+XH) in blue, of crude plus condensates plus NGPLs in purple, and all liquids in green. Comparison with that of crude oil plus condensates in brown and all liquids in orange in the United States. For the United States, the Figure shows the peak of the US Lower 48 in 1970 (predicted by Hubbert in 1956), the bump of the Alaskan discovery with its peak in 1984 and the rapid rise of the LTO from 2010. Source: J. Laherrère (2016).*

LTO, as well as that of liquids (condensates and NGPLs) extracted from their shales (source-rocks). But LTO production currently accounts for only a modest share of all liquids oil worldwide, about 4.4 Mb / d in 2014 (Tables 7 and 13). On the other hand, the agencies do not foresee a dramatic increase in the LTO in the United States, and even predict its decline around 2020–2030 (Figure 59). In fact, as can be seen clearly in Figure 44, it is mainly the production of conventional crude oil that the IEA accounts for the increase in the production of all liquids petroleum. This is in contradiction with the decline in conventional production observed since 2005 (Figure 58), and the rapid decline in its 2P reserves (Figure 35)!

All this seems to indicate the very near future of a peak or a plateau of oil all liquids world-wide between 95 and 100 Mb/d, in accordance with the predictions of ASPO and not those of the agencies, except if, as will be discussed later, the American success story of the LTO was spreading rapidly on a global scale. But it still does not take the way.

147

As for the energy content of these products, a recent study by X. Chavanne 2016 (Figure 61) on the production of natural oils (i.e. excluding synfuels and refinery gains) between 2000 and 2015 still shows an increase in HHV (in Mboe/j as indicated in Chapter 2.2) for the production of C+C+LTO+XH+NGPLs. However, this author anticipates its gradual decline by 2015. As for conventional oil (C+C), a slow decrease is observed from 2005 on.

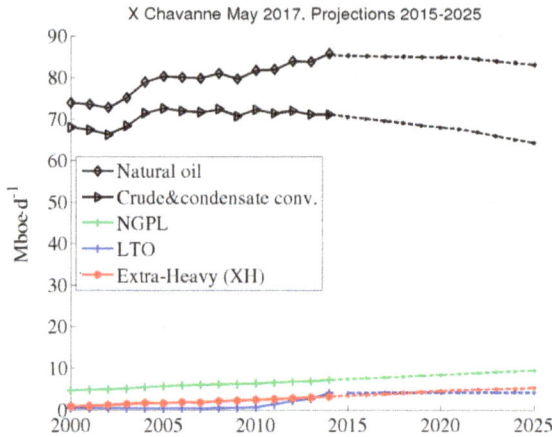

Figure 61 Evolution of energy quantities (High Heating Value (HHV), in Mboe/d) contained in natural oils (thus excluding synfuels and refinery gains) and their components from 2000 to 2015, and projections for the period 2015 to 2025, according to Xavier Chavanne (2016).

Table 13 shows the energy value, in Mboe/d, of the quantities produced from the natural oil constituents in 2014 and their comparison with these quantities expressed in volume (Mb/d).

Table 13 Quantity of energy (HHV, Mboe/d) contained in the components of the 2014 production of all liquids petroleum, according to Xavier Chavanne (2016), and comparison with the volume quantities (Mb/d) according to the EIA.

Nature of component	Mb/d	Percentage of total	Mboe/d	Percentage of total
Conventional + condensates (C+C)	70.3	75.4	71	81.6
Light tight Oil (LTO)	4.4	4.7	4.1	4.7
Extra-Heavy (XH)	3.2	3.4	3.1	3.6
NGPL	9.8	10.5	7.1	8.2
Biofuels (agrofuels)	2.5	2.7	1.3	1.5
GTL+CTL	0.5	0.6	0.4	0.4
Refinery gains	2.5	2.7	0	0
All liquids total	93.2	100	87	100

The amount of energy made available to the world society by natural oils would thus be already decreasing, and so would all liquids petroleum, given the low energy contribution of synfuels and their insignificant growth. The corresponding amount of energy per capita, in view of the increase in world population, is certainly decreasing. It should be noted that the increasing share of output devoted to this production, i.e. the reduction of energy return rates (EROI) to production (see Chapter 2.2), further reduces the amount of energy thus made really available to the world society.

It should also be noted that, concerning natural oils, i.e. excluding synfuels and refinery gains, C+C category, that is to say conventional oil, represented in 2014 80.2% of the total in Mb/d, and 83.2% of the total energy content, in Mboe/d.

LTO's production may be higher than that foreseen by ASPO, or even by agencies, if extraction technologies upgrade significantly and especially if other nations than the United States start an important production. It is currently the subject of passionate discussions. But we still do not see the development on a world scale announced: the countries most talked about for a large production of oil (and gas) from source-rocks are for the moment Argentina (Vaca Muerta in the Neuquen basin) and Russia (Bazhenov formation in the Western Siberian Basin). China and Australia are also included. Saudi Arabia and Iran could be interested. But only Argentina has achieved some success.

It should be noted, however, that oil production of this type has a high resilience due to its high speed of implementation. While it takes 5 to 10 years to start producing oil from a conventional deposit after it is discovered, it only takes a few months to start producing oil, or gas, from source-rock. But at present, given the low price of oil, the profitability of shale oil production has become insufficient for new drilling to be made, except in the best sweet spots. Then, the production of a well of this type decreases by about 90% in 3 years (see Appendix 2). The decrease of new boreholes therefore reduces overall production very rapidly. But a rebound in the price of oil clearly above the break-even, that is to say the selling price above which the producer no longer loses money, would revive it very quickly in the United States. On the other hand, with the decline of wells being less and less rapid over time, production could stabilize at a relatively high level once a very large number of wells have been drilled.

Ultimately, the future of all liquids oil production depends mainly on two major players: the production of conventional crude oil, in decline, and the production of shale (source-rock) oil, which until 2015 has been increasing very much in North America.

It is however currently stagnating and does not show any clear sign of the possibility of significant development on a global scale except in Argentina. The obstacles to this development are not only geological, but also politico-economic.

The question of the future development of shale oil is therefore still open. But by 2040 production outside the United States would need to be equivalent to at least 4 times the country's current production, to offset the decline in predictable conventional crude oil by that date. Such a development on a world scale would be truly prodigious!

A peak, at best a stagnation for a few more years, of all liquids oil production is therefore very likely in few years from now. For ASPO, this peak would take place around 2020.

The natural gases

As with petroleum, production is evaluated in volume, under normal conditions (15 °C, one atmosphere). The units used in the Anglo-Saxon world and more generally by the oil companies are the trillion (10^{12} = Tera) of cubic feet (Tcf), whose energy equivalent in oil equivalent is about 24 Mtoe, and the billion (10^9 = Giga) of cubic feet (bcf), the equivalent of which is therefore 0.024 Mtoe. The International Energy Agency (IEA) uses the billion of m^3 ($G.m^3$, km^3), which is about 0.9 Mtoe.

Production in 2014 was roughly equivalent to 3 billion of toe (Gtoe). Recall that this is the marketed dry gas, i.e. gross (gross) production deprived of the condensates and the NPGLS it initially contained, and less the quantities reinjected in the deposit or flared with a torch. Recovered condensates and NGPLs are accounted for as we have seen in all liquids oil.

This production consists of gas from conventional deposits, and from unconventional deposits: shale gas, tight gas, coal seam (CBM) and coal mine (CMM). The production of CMM is anecdotal. Syngas production is not significant. Unconventional gas world production, in oil equivalent, was in 2013, according to the IEA, about 280 Mtoe for shale gas, 200 Mtoe for tight gas and 60 Mtoe for CBM (Table 14), i.e. 540 Mtoe in total. The main part, about 400 Mtoe, was produced by the United States. This production consisted of about 244 Mtoe for shale gas, 115 Mtoe for tight gas and 40 Mtoe for CBM.

Table 14 *Unconventional gas production in the world and the United States in 2013, in million tonnes oil equivalent (Mtoe). Source IEA.*

Nature of gas	World	USA
Shale gas	284	244
Tight gas	200	115
Coal bed gas (CBM)	58	40
Total	542	399

But tight gas and CBM are in the United States produced from anciently operated deposits, and are in fact in decline, since 2008 for tight gas and since 2010 for the CBM (Figure 62). There remains the shale gas, which has so far developed very rapidly, especially after 2005, but for which it is very difficult to predict what will happen next. While production still increased much in 2014 and 2015, a small

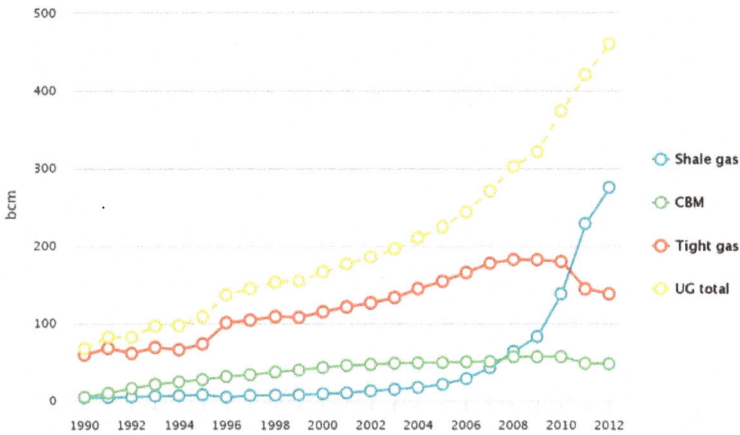

Figure 62 *Unconventional gas production in the United States from 1990 to 2012, in billions of m³ (bcm, 1 bcm is about 0.9 Mtoe). We note the very rapid growth of the shale gas after 2005, but also the peak of tight gas in 2008 and that of the Coal Bed Methane (CBM) in 2010. Source: IEA (http:/www.iea.org/ugforum/ugd/united%20states/).*

decline has been observed in 2016 (see Figure 53) and confirmed in 2017. This is said to be due to the collapse in market oil price after mid-2014. Indeed a large part of shale gas output was profitable thanks to their content in condensates and NGPL, sold at the price of oil. Some believe that the previous very rapid development will nevertheless resume as soon as prices of oil will increase again, and also thanks to technological progress making its "economic model" more and more efficient. Others, such as J.D. Hugues already quoted, foresee on the other hand an upcoming peak of shale gas in the United States at about the same time as that of the shale oil, i.e. before 2020.

It should be noted that in the United States, the world's largest producer of gas at the present time, gas prices are not linked to the price of oil as in the rest of the world (see Chapter 4), but fixed on the American spot market (Henry Hub). The evolution of its production is thus an essentially internal affair.

On a global scale, it can be seen (Table 14) that if the United States carries the lion's share for shale gas, it is less so for tight gas, where there are other producers, mainly Canada, followed somewhat behind by Russia. Again, these are anciently operated deposits, and it is doubtful that there will be a significant increase in this type of production worldwide in the years to come.

In the CBM, non-US coal-producing countries have small productions, mainly China, Australia and Canada. It is not impossible that, on a global scale, these products will develop significantly, but it is doubtful whether this will be a very important contribution to global gas production in the future.

There remains the shale gas, of which one cannot exclude that it could experience a very strong development at the world scale, like that happened in the United States. But there is no sign for the moment of the worldwide launching of an "American success story".

Figure 57 shows the peak of natural gas around 2030, at about 3.6 Gtoe. There is no obvious reason for changing this date at this time. However, the situation here appears to be somewhat more open than for all liquids oil, the main component of which, conventional crude oil and condensates, has already begun to decline and for which the production of shale oil should increase spectacularly to compensate this decline, but it is not yet at the rendezvous.

The Coals

The evolution of coal production is also difficult to predict, due to the lack of rigor in the definition of reserves, but also to unreliable historical production data in some major producer countries, especially China, India and Indonesia. This makes predictions from production histories very uncertain, and it should be noted the speed at this time of the development of the countries of Southeast Asia, India, Indonesia, Vietnam... which is actually pledged on their consumption of coal! In fact, China, the world's largest producer, has peaked in 2013, and this date might also mark the peak coal at the global scale, as suggested on Figure 57.

Various

With regard to the oils produced by the pyrolysis of bituminous shales, to the underground gasification of coal and bituminous shales, to gas hydrates and gases from deep aquifers, much cannot be said at the present time. Technological improvements and price increases may eventually make them profitable, but their implementation difficulties and environmental disadvantages will constitute a major brake on significant development, during this century they will probably still have as they have today the status of resources of the future.

It should be noted that the ASPO models are based on geological constraints (underground), with no political-economic constraints (above ground) other than the constraints "Business as usual" (BAU). In the event of a large forced decline in production as a result of an economic or political crisis, the peaks will be delayed and their height lower, the only thing remaining about constant being the area under the production curve once the latter definitely arrested, that is, the ultimate. For example, the oil shocks of 1973 and 1979, leading to a deceleration in world consumption, significantly delayed the peak oil date.

It is with oil that the deadlines are the closest and the best understood, and it would be good for the "policy-makers" of the industrialized countries to think about it very seriously, given its extreme importance in the economies of these countries.

4

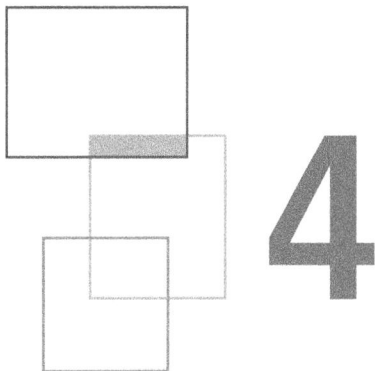

A little economics: fossil fuel prices

The economics of fossil fuels is not the subject of this book, but it is difficult not to say a few words, particularly as regards the price of oil, to which the consumer is of course very sensitive.

4.1 Oil prices

The most popular "law" of the economy is that of supply and demand, according to which on a "free market" the price of a product increases when its supply is lower than its demand, and decreases in the opposite case.

A very sharp fall in oil prices was observed in 2014–2015, following a very significant increase from 2003 to 2014. A common explanation is that, as in the period of the oil shocks of 1973 and 1979, very high prices have encouraged investment in exploration and production, the search for oil substitutes, and a slowdown in consumption: investments are now bearing fruit, and in particular North American LTO and Canadian bitumens arrived in force on the market, supply grew above demand. This caused a fall in prices and a slowdown in investment. One would thus be in an oil countershock, similar to that which in 1986 had succeeded to the oil shocks. With lower prices leading to an increase in consumption and a decline in investment in exploration and production, surpluses are expected to be absorbed and prices to rise again in the future. This phenomenon is not limited to petroleum, but concerns many other raw materials: this is called the commodities cycle.

The decline would be amplified by the behavior of stock market operators, who prefer to seek more profitable investments. Paradoxically, this would also be due to an increase in production in those producing countries for which oil export income is an essential part of their budget and which compete with other producers to minimize their losses. The same holds true for companies operating production-sharing fields, for which, when the price is low, production must be increased to meet the financial contractual commitments (see Chapter 3.1.1 and note 27).

The decision by the OPEC leader, Saudi Arabia, not to support prices by reducing production (see below), as it did after the oil shocks, but at its expense, as well as in 2009 but then without any damage for it, when the oil price fell sharply following the outbreak of the 2008 financial " subprimes" crisis, would have triggered this bearish cycle. Some also see the effect of major geopolitical maneuvers in this period of severe disorder in the Middle East and of growing confrontation between the United States and Russia. The decision of Saudi Arabia, announced by commentators as a necessity for it not to lose its share of the oil market in the face of competition from North American LTO, would have been dictated in reality by the States which would have chosen in this favorable context of surplus production relative to demand, to weaken Russia and other producing countries such as Venezuela, even if it means sacrificing their producers of LTO.

Let us not lose sight of the fact that the supply of oil has long been regulated by a quota policy: initially, this policy was the result of a cartel of the major Western oil companies, following the agreement of Achnacarry[36] in September 1928.

Then the major exporting countries took over, following the creation in 1960 of the Organization of Petroleum Exporting Countries (OPEC), which also functions as a cartel, with the allocation of production quotas to its countries. The strength of this cartel is that it can set prices on the world market by restricting its supply because other producing countries, known as NOPEP countries, do not have sufficient production capacity reserves (moreover less and less) to compensate by bringing quickly their production up. For example, OPEC reduced its output by 4.2 Mb/d in January 2009 (Figure 63) to drive up prices. From these countries, Saudi Arabia has played a leading role as a "swing producer", mainly in terms of its political and financial interests, thanks to its ability to reduce its production to maintain or raise prices, or to increase it to bring prices down. This capacity for rapid increase in production is what is called "spare capacity" (Figure 64).

---

[36] The main provisions of this agreement concluded at the Castle of Achnacarry in Scotland between the 7 main oil companies of the time (the 7 sisters), all Anglo-Saxon, were a stabilization of their market shares at the level of 1928, and the elimination of possible excess production at the level of fields or refineries. A cartel of companies was then created, which no longer competed with one another.

154

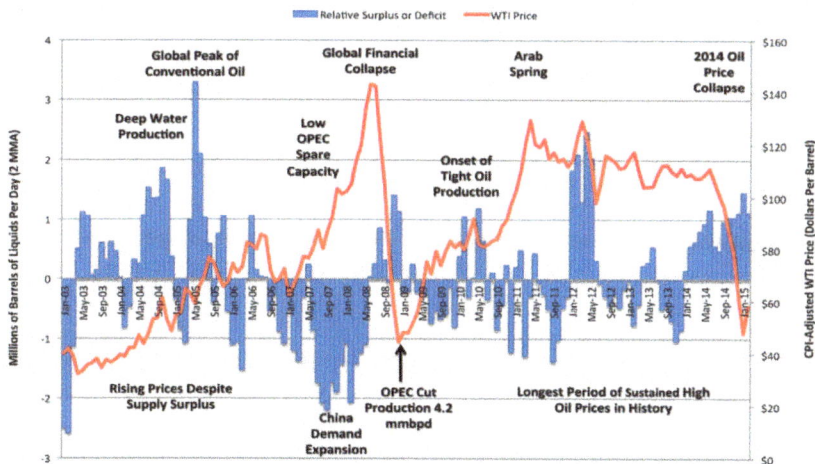

Figure 63 | *Comparative evolution of WTI prices in USD per barrel on the US market (New York Mercantile Exchange, Nymex) (red curve) for the period 2003–2015, and differences between supply and demand of all liquids oil, in millions of barrels per day (blue histogram). Source Art Berman.*

Figure 64 | *Comparative trends in the value of the euro to dollar value and WTI oil spot price from 2006 to 2016, and successive QEs and OPEC interventions. In red, the QEs of the Federal Reserve System (Fed) of the United States, in blue those of the European Central Bank (ECB). According to Art Berman (personal communication), the correlation coefficient between WTI price and dollar value is 0.96 from 2014 to 2016! Source: Courtesy J. Laherrère and N. Meilhan.*

By the end of 2014, Saudi Arabia did not want to play again this role, contrary to the expectations of the stock market operators, and chose to increase its production rather than reduce it.

Saudi Arabia appears to be the only country in the world that still has such a spare capacity. For how much longer? It is difficult to know because it is one of those countries where information on oil fields is very difficult to access.

It should also be noted that the volume of trade in commodity markets increased explosively after 2001, the date of China's entry into the World Trade Organization (WTO) and the beginning of the very rapid growth of its economy, and therefore of its raw material needs. The considerable increase in transactions, and the economic health of China, thus also appear to have become major factors in the instability of oil prices, and more generally of commodity prices. China's fantastic growth slowed in 2014 and 2015, causing a slowdown in demand for raw materials worldwide. Yet China's demand for oil has continued to grow as a result of the rapid development of its vehicles fleet, with 20 million more vehicles put into circulation every year!

But the current fluctuations in oil prices may have other causes than those mentioned above:

Figure 63 is due to Art Berman, a well-listened American analyst who compared the price changes from 2003 to 2013 of West Texas Intermediate (WTI), which is the benchmark oil quality in the US market (on the European market is Brent), with variations in the gap between supply and demand on the world market. This period is that of the history of the oil that has known the most long-lasting very high prices.

There is no obvious correlation between the two, except for the sharp price changes from 2007 to 2009 and 2014 to 2015. It is also noted that the magnitude of price fluctuations is much higher in percentage than of the gap between supply and demand, which remains essentially within a range of ±2 Mb/d, i.e. roughly $\pm2\%$ of world production. This range is about the same as the margin of uncertainty about the actual amount of this production (Figure 42). This suggests that these short-term price movements are not really the result of objective information from stock market operators on the real availability of oil, which we can see that world production, notwithstanding these uncertainties, varied relatively little over the recent period (Figure 60).

On the other hand, during this period, there is a very good anti-correlation between the price of oil and the monetary Dollar Index, that is to say the value of the dollar against a basket of the main world currencies used in trade. The price of oil falls when the dollar index rises, and vice versa. The strong correlation between the price of oil and the ratio of the value of the euro to the value of the dollar confirms this relationship with monetary exchange rates.

The Dollar Index was created in 1973 by the United States Federal Reserve (Fed) following the abandonment of the gold standard in 1971, under the presidency of Richard Nixon. The Dollar Index is an index of the economic health of the United States, but perhaps it could be said of their stock market health as compared to that of the nations issuing other currencies, and it is indeed well correlated with the principal US stock market index, the Dow Jones. An interpretation commonly advanced is that the price of oil is valued in dollars, while consumers outside the United States react to the local currency price. When the Dollar Index rises, it takes more local currency to buy the dollars needed to buy oil on the market, which curbs demand and lowers prices. The opposite is true when the Dollar Index falls.

But this does not explain the fluctuations of the Dollar Index from 2003 to now, nor that of the ratio between the value of the euro and the value of the dollar!

Art Berman, quoted above, interprets the sharp drop in the oil price of 2014–2015 by the combination of a large surplus of supply on demand and the end of quantitative easing (QE) in the United States, that is to say, of monetary creation (equivalent to the old money printing press) in order to support an economic activity very weakened by the crisis of 2008, and thus resulting in a better health of the US economy. The successive QEs, by the influx of dollars they had caused on the money markets, would have lowered the dollar and thus the Dollar Index. Their interruption mid-2014 would have provoked the opposite phenomenon. The Eurozone has taken over, in turn doing quantitative easing, making the dollar even higher by a weakening of the euro.

This interpretation seems to be borne out by the examination of Figure 64, where the WTI price changes and the value of the euro to dollar value from 2006 to 2016 are compared.

In this Figure, we clearly see the relationship between the quantitative easing policy of the US Federal Bank (Fed), the euro/dollar ratio and the oil price. We can also see the influence of OPEC policy, which pushed up prices in 2009 following the sharp fall due to the economic crisis of 2008 and reduced them at the end of 2014 by its decision not to counter the fall of prices that would have been initiated by the end of quantitative easing.

However, the QEs and OPEC interventions do not explain the developments observed from 2002 to 2008. A possible interpretation is that from that date the Fed issued an abundance of "petrodollars" to support the petroleum activity. But with the first release of the euro in 2002, it began to compete with the dollar as a reserve currency: some of these petrodollars were converted into euros, thus raising the value of the euro to the dollar. Then, the relay would have been taken by the considerable increase in Chinese demand, at a time when the spare capacity of OPEC was at the lowest (Figure 63), and until the economic crisis of 2008 which made brutal lower demand.

But there is another interpretation which, on the contrary, closely depends on oil prices, essentially on the day-to-day behavior of the stock market. This interpretation implies, therefore, that the price of oil governs exchange rates on the foreign exchange market (Forex) and, by the need for adjustment, the monetary policies of the major central banks. Oil prices are predominantly the result of fluctuations in supply-demand differentials, although these differences are relatively small and of the size of the stocks in the major consumer countries. In this sense, the rapid development of shale oil in the United States in a period of stagnant demand would be the main cause of the current situation; Let us observe that if, according to the first interpretation, money flows govern oil flows, the second is exactly the reverse: it is the oil flows on the markets which, through the fluctuations of supply and demand and the size of stocks, and because of the predominant share they represent in terms of market value, govern the flows of money and, consequently, the global economy.

At this time scale, it is therefore very difficult to distinguish in fluctuations in the price of oil what is due to politico-economic fluctuations (an economic equivalent of Brownian motion in physics?), to monetary exchange rates, to stock market behavior, and to physical availability, although the latter is necessarily one of the determinants. Predicting them even in the medium term is a divinatory exercise, and those who have risked it, such as the International Energy Agency (IEA) (Figure 65), have so far verified it at their expense.

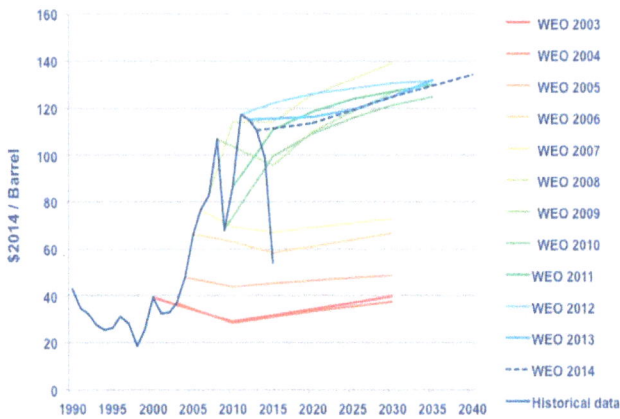

Figure 65   *Forecasts of oil prices in 2014 dollars per barrel, made by the IEA in its successive WEOs from 2003 to 2014, and comparison with reality. Source Carbone 4, Courtesy J.M. Jancovici.*

However, it seems *(Caruana, 2016)* that stimulating economies by financial techniques plays the greatest role in recent developments: after the first oil shock in 1973, there has been an increasing public and private debts to finance oil exploration and production, and to boost and sustain growth, first in Western countries and then in so-called emerging countries. Then after the economic crisis of 2008, the major central banks flooded the liquidity markets with the quantitative easing method and lowered their interest rates to encourage credit, and thus consumption and economic activity. All this ultimately led to very little. It would seem, therefore, that the financial economy has less and less hold on the real economy, that is to say, production of material goods and services, because it encounters physical limits and for this reason is increasingly pushing. It is probably in this increasingly marked

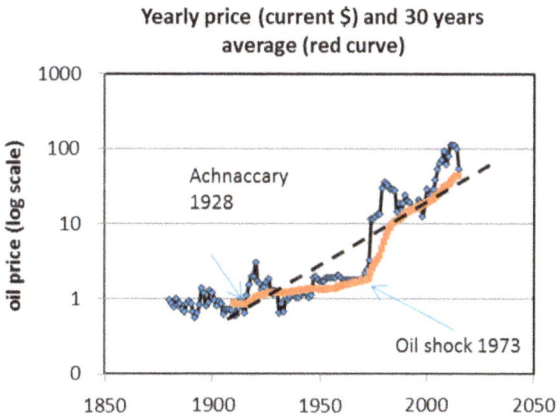

Yearly price (current $) and 30 years average (red curve)

Figure 66   *A possible trend in oil prices in dollar terms over the 30-year period, which corresponds to the average half-life of large oil deposits. The period before the oil shocks resulted in an agreement on prices between the main Western oil companies of the time, decided at Achnacarry in Scotland in 1928. This period ended with the creation of OPEC in 1960. Note that they are current prices and therefore not adjusted for inflation, and that they are indicated on a logarithmic scale. This chronicle goes until 2015. Courtesy Paul Alba.*

disconnecting between monetary economy and physical economy that the greatest source of instability is found, not only in the oil sector.

Over longer periods, it is possible to observe some trends. The graph in Figure 66 shows the evolution of "filtered" prices, i.e. after eliminating background noise from short-term fluctuations and then averaged over 30-year periods, which corresponds approximately to the half-life of a large oil field. There are then two major historical trends: the first is the takeover of the major Anglo-Saxon oil companies in the markets, formalized by the Achnacarry agreement: it is characterized by slow average growth and relatively low fluctuations in dollar prices currents. It reflects the fact that the oil companies then transferred their profits from the Upstream (Exploration-Production) to Downstream (Refining-Distribution).

The second one begins after the creation of OPEC (OPEC), which this time corresponds to the control of the major exporting countries on the markets. The underlying trend is then a rapid increase in current dollars, but with enormous fluctuations. This can be explained by the fact that the discoveries of oil occurred during this period with decreasing average yield.

The choice of current dollars to monitor oil prices rather than constant dollars in a recent year is justified by the fact that the economy operates with current dollars and that converting them into constant dollars with discounting coefficients are based on often questionable assumptions.

In a few years, the global supply should, as we have seen, peak and then gradually decline for geological reasons, on which the economy and politics will have less and

159

less hold. As a result of these decreasing returns from oil exploration and production, it would seem logical that the underlying trend would then be a sharp increase over time. This trend can also be reinforced by the development of an energy policy guided by environmental concerns. However, prices will not follow this trend if a decline in demand, whether forced by a lasting global economic crisis, voluntary by economic measures, or due to the large-scale development of substituents for petroleum in its main uses (mobility, heating, petrochemicals, electricity) accompanies this decrease in supply. The price of oil will doubtless remain very uncertain in the years to come, and repeated crises are to be expected.

Over the past 25 years, world oil production has fluctuated much less than the price of a barrel.

4.2 Gas prices

Figure 67 compares the evolution of the price of oil (Brent) with that of gas prices on the basis of the same unit of energy content, the British Thermal Unit (BTU), an Anglo-Saxon unit commonly used by markets to assess gas quantities.

Gas prices are very different depending on the market. This is due in the first place to the very high cost of transporting gas (by gas pipelines or LNG tankers), as markets cannot have supplies at competitive prices beyond a certain transport distance from production sites. The prices on a market are all the higher when its sources of supply are distant. This is a big difference with oil, whose transport costs are much lower in comparison, and therefore prices are very similar from one market to another: the difference between the two main references, the WTI for the US market and the Brent for the European market, is currently only a few percentage. It is also observed that, for the same amount of energy contained, gas has been on average less valued than oil, and this phenomenon has been accentuated during the recent period of very sharp increases in oil prices from 2003 to 2014. This is due to their difference in physical properties, a gas having, compared with a liquid, the disadvantage of requiring more complex and costly installations (compression-decompression, transport, storage, etc.) for the users.

Finally, except for the US and Canada markets after 2009, there is an overall correlation between gas prices and the price of oil. This is due to the particularities of contracts between gas suppliers and consumers, in which prices are largely indexed to oil prices. The price of oil is therefore a guiding price for gas prices. However, it has become less and less true in recent years with the gradual increase in the share of gas that escapes contracts and is evaluated directly on the markets.

In the United States, since the shale gas boom began in 2005, the price of gas is no longer indexed to that of oil but quoted on the markets.

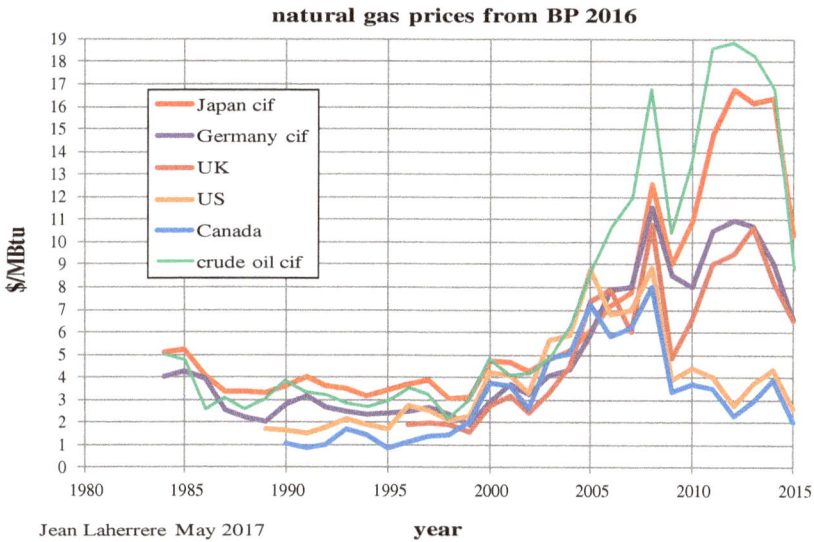

natural gas prices from BP 2016

Japan cif
Germany cif
UK
US
Canada
crude oil cif

Jean Laherrere May 2017   year

Figure 67   *Variation of gas prices in different markets from 1984 to 2015 and comparison with the price of oil (Brent) for the same unit of energy content, in this case the million of British thermal unit (MBtu), equivalent to 18 barrels, 1055 MJ, or 0.025 toe. There is a strong overall correlation between oil and gas prices, except for the United States and Canada after 2009. The very high price in Japan comes from the impossibility of supplying it other than by liquefied natural gas transported by LNG tankers. Cif means cost, insurance and freight, that is, the sum of the selling price, insurance and transportation costs.*

4.3   Coal prices

The price of coal is valued in dollars per tonne. For a long time coal was consumed mainly in its producing country. This is particularly true of lignite, which is consumed very largely at the very place of its production to produce electricity. But the hard coal was already the subject of an important international trade in the 19th century. The oil shocks of 1973 and 1979 have revived its momentum *(Martin-Amouroux, 2008)* and in recent years as well, as a result of China's strong demand and the search for foreign markets by the United States. Exported coal, however, currently accounts for only about 20% of the world's total coal output, compared to just under 30% for gas and about 50% for oil, according to the BP Statistical Review.

The cost of transport is higher than for oil, but less than for gas, and thus influences prices less than in this case. However, there are several markets.

These prices vary according to qualities and markets, but there is a general correlation from one market to another (Figure 68).

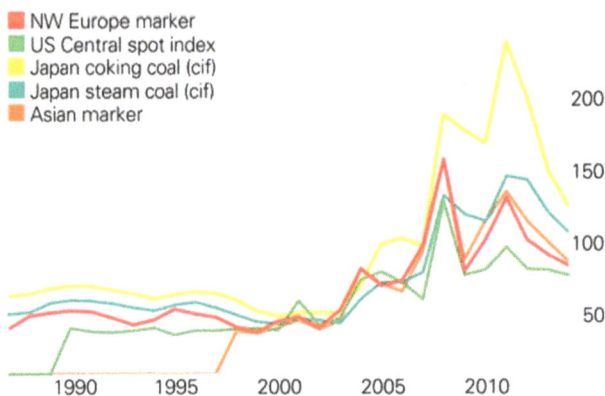

Figure 68 *Recent trends in coal prices in major markets, in dollars per tonne. Prices continued to decline after 2014, as for most other commodities denominated in dollars. Cif means cost, insurance and freight (selling price + insurance + transport cost). With the exception of coking coal in Japan, what is shown here is steam-coal, the average energy content of which can be estimated at about 0.6 toe/t. Lignite is not traded internationally. Source: BP statistical review 2015.*

Again, price movements have long been driven by oil prices. But after 2011, there is a sharp decline in prices, and for steam coal, mainly used for electricity production, tightening in a range of $ 70–100 per tonne in 2014, depending on markets. This corresponds to about $ 120 to $ 170 per toe ($ 16 to $ 22 per barrel), or to make these prices comparable to those in Figure 67, to $ 3-4 per million of BTU. It is currently the cheapest of fossil fuels by quantity of energy contained, with the exception of the US market where gas is cheaper, especially for electricity generation, a gas plant being much less demanding in investment than a coal-fired power plant. Since supplier-to-consumer contracts are less rigid than for gas, it is also the most flexible energy that can adapt quickly to changes in the market.

Two consequences for the United States are, on the one hand, the replacement of coal by gas in the production of electricity and, on the other hand, the increase in their exports of coal to countries where the price of gas is much higher than that of coal, Europe, Japan… where, conversely, by a "game of musical chairs" it is therefore imported coal that replaces gas in the production of electricity.

These were unexpected collateral effects of the shale gas boom in the United States.

4.4 Trends in market shares of primary energies

Despite the various subsidies granted by governments to individual energy production according to their energy policy, energy market prices drive for the essential investments in their productions.

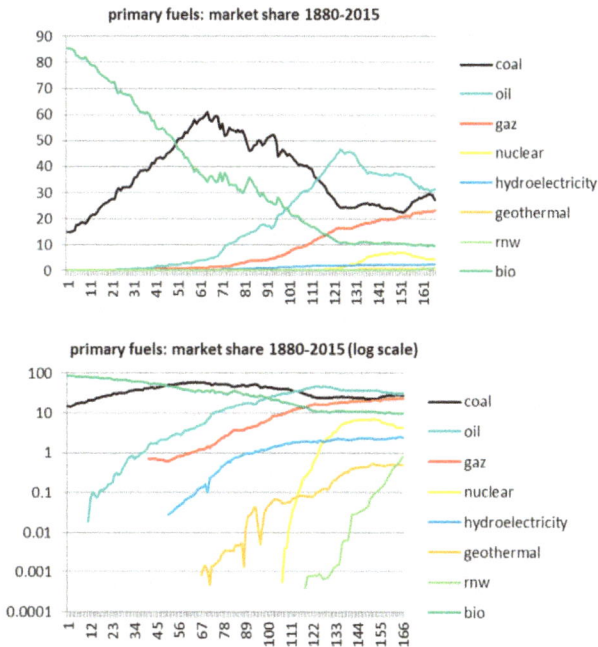

primary fuels: market share 1880-2015

primary fuels: market share 1880-2015 (log scale)

Figure 69 *Evolution of world market shares of the main sources of primary energy from 1850 to 2015: The label rnw covers solar thermal, solar photovoltaic and wind. Above, in percentages, below in logarithm of the percentages. The Figure below shows better the evolution of market shares of "very small players", rnw, geothermal. Courtesy Paul Alba.*

Each primary energy source has a market share based on its physical characteristics and its cost of production and transportation.

The chart in Figure 69 is a chronicle of the evolution of the market shares of the various sources of primary energy since 1850. It shows the overwhelming preponderance of fossil fuel and biomass (mainly firewood). Within this group, there is a breakdown in developments after the first oil shock in 1973: considerable decline in the share of oil in favor of gas and coal, stabilization of the share of biomass. This is the time when oil is gradually being abandoned for the production of electricity, to the benefit of coal, gas and nuclear power. From 2001, when China, the world's largest producer and consumer of coal, enters the world trade organization and its big economic takeoff, there is another break: new oil decline and stabilization of gas, to the benefit of coal. However, for a few years now, coal is loosing market shares. This is due mainly to a decline in the coal production of China, and to the replacement of coal by gas for the production of electricity in the US.

The other sources of energy, nuclear, hydro, wind, solar, geothermal with about 10% in total market shares currently, are here small players. But above all the investments that would have to be made to develop these sources so that they become a majority are huge compared to their present importance! Indeed, current world primary energy production is of the order of 14 Gtoe today. Earning 1% market share for an energy source means producing 140 Mtoe more. By virtue of what might be called a mass action law applied to the economy, it is relatively easy to do for the big players, where there is already enormous capital invested and the infrastructure available, and for which this does not represent very high annual growth rates. Coal during its initial take-off, took only about 1% per year against wood, and again only takes about 1% a year against oil and gas from 2001 to 2013. Doing the same for very small players[37] would represent phenomenal growth rates. Without gigantic aid from governments or very high taxes on fossil fuels, markets will not react spontaneously, and it would be over a very long period.

This difficulty appears even more clearly in the diagram in Figure 70 due to Olivier Rech, where the year-over-year change in the market share of each source is plotted against the amount of energy it produced in that year, in ktoe. The abyss that separates renewable energies and even nuclear from fossil fuels becomes very visible, and coal has been currently so far the big winner of the current competition.

The essential role of fossil fuels in the energy transition also becomes very visible: By physical necessity, it is the evolution of their production possibilities it will depend on essentially, and it is the speed of this evolution which will give the tempo.

For now, we are witnessing not the triumph of renewable energies or nuclear energy, but the great return of coal (*Martin-Amouroux, 2008*). However, this will not be very durable, given the geological limits to its production.

But, let us recall it tirelessly because it is generally far from being clearly perceived, the quantity of energy made available to human societies is much more important because it has a direct physical relationship with the quantity of goods and services that can be produced, than its price. And there is a close correlation between primary energy consumption, and especially oil, and GDP in the world, above all in the industrialized countries.

Very few economists currently seem to clearly understand the meaning of this relationship, which is intuitive for a physicist or an engineer. Some have nevertheless been very far away in reflections on this subject, such as *Nicholas Georgescu-Roegen (1971)*, and *Rainer Kümmel (2011)*, for which human societies are structured by

---

[37] It should also be remembered that the physical characteristics of these small players are currently powerful obstacles to their development. To put it briefly: Hydropower is handicapped by limited resources and in many cases by its remoteness from the major centers of consumption, apart for some very well-endowed countries. Nuclear power is hampered by resources which, while awaiting the development of breeder reactors, are currently too uncertain, and by fears, justified or not, about its safety. Wind and photovoltaic solar energy are handicapped by their fatal nature (their production depends on meteorology, not on human will) and their intermittency, which oblige when they are lacking, in the absence of massive and cheap storage of electricity, to use other sources of energy. Deep geothermal energy is limited by the low geothermal flux. It should also be noted that none of these energies can easily replace fossil fuels and especially petroleum in their main uses, except for hydraulics and nuclear power, which can easily replace gas and coal in electricity production.

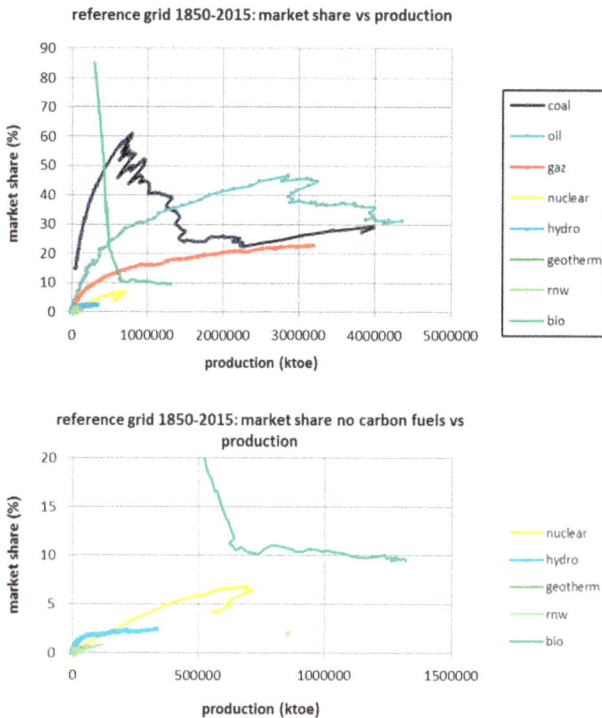

reference grid 1850-2015: market share vs production

reference grid 1850-2015: market share no carbon fuels vs production

Figure 70 *Rech diagram: on the ordinate, the successive annual market shares from 1850 to 2015 of the various primary energy sources, in percentage, and on the x-axis their production on the same dates, in ktoe. Above, for all sources of primary energy, below for no carbon sources only. Courtesy Paul Alba.*

energy flows, and *Robert Ayres (1998)*, who proposes to establish balance sheets in joules rather than in dollars.

But the overwhelming majority of models used by current economists want to consider energy only by the book value of its consumption, not by the quantities used. A few, however, have expressed concern, such as *Joseph Stiglitz (1974)*, who proposed an economic model that included a factor related to the availability of energy resources to predict changes in GDP, but in an abstract way, without any particular idea about the evolution of this factor in the future. Another example is *Gaël Giraud and Zeynep Kahraman (2014)*, who have shown, from the study of the economic performance of 33 countries, the determining role of their amount of energy consumed and the underestimation of this role in current economic models. The economic models linking the amounts of energy consumed to GDP are still to be written.

It is therefore to the evolution of the available quantities of energy, starting with that of petroleum, that we must take priority, rather than prices, of which we have seen are largely unpredictable.

5

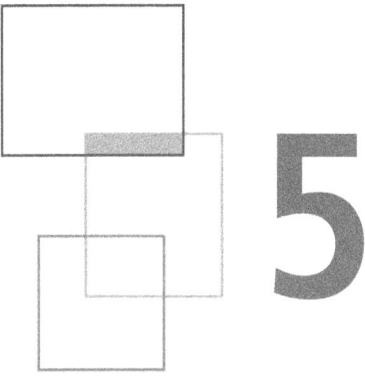

Fossil fuels and climate

According to the Intergovernmental Panel on Climate Change (IPCC), anthropogenic CO_2 emissions are the main contributors to radiative forcing[38] leading to an increase in the temperature of the Earth' surface currently observed. Then, 85% by mass of these emissions are now the result of the use of fossil fuels, the remainder being due mainly to changes in land use (mainly deforestation) and to a lesser extent to the use of carbonates for the manufacture of cement.

---

[38] Earth can only exchange energy with space around it in the form of radiation. It receives radiative energy from the sun light, about 240 W/m² of its surface on average, and reflects about 30% of this energy without altering the light spectrum, this is called its albedo. The remaining 70% is absorbed by the atmosphere and the earth's surface, increasing its temperature and thus producing infrared light radiation. When the energy of the solar radiation coming from the space is balanced by the albedo plus the energy of the infrared radiation leaving towards space, which was still the case before the Industrial Revolution, it is said that there is radiative equilibrium. The radiative forcing is a disturbance of the radiative exchanges (solar / terrestrial infrared + albedo) induced by a modification, anthropic or natural. For example, a sustained increase in the energy of the solar radiation received by the Earth globe will heat up the atmosphere and the surface and cause an increase in the infrared radiation of them, so as to regain a radiative balance between the earth and space. But the temperatures of the Earth's surface and atmosphere will remain higher than before, as long as the increase in the power of solar radiation persists. An increase in the greenhouse gas content of the atmosphere produces the same result by returning to the ground's surface a part of the emitted infrared radiation. These radiative forcings are expressed with the same unit as the power of the received solar radiation and the infrared radiation emitted, that is the W/m² of terrestrial surface.

It is easy to calculate from the curves in Figure 57 the associated carbon dioxide emissions (Figure 71).

world fossil fuels CO₂ emissions & production in toe

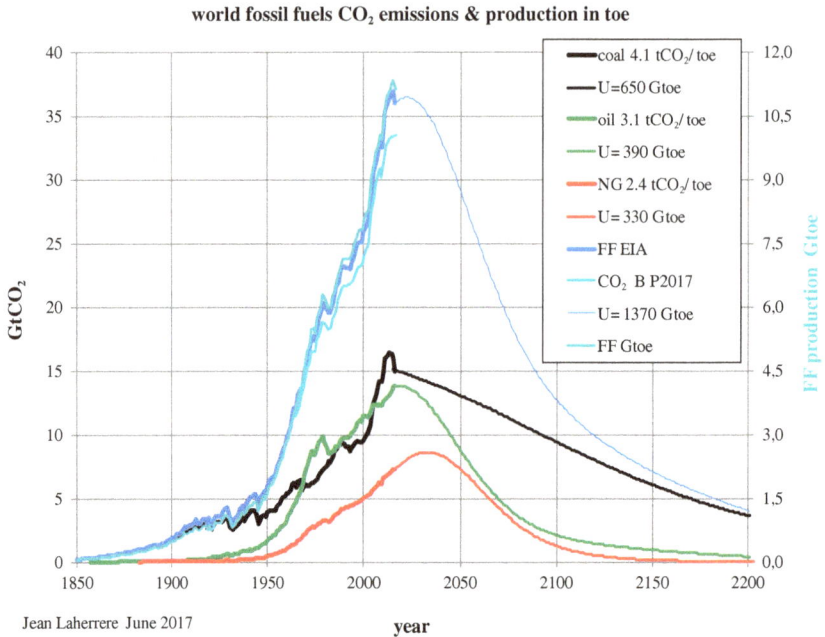

Jean Laherrere June 2017

Figure 71 CO_2 emissions from coal, oil and gas, and total fossil fuels (FF), and forecasts up to 2200, in $GtCO_2$. The equivalences taken are, in tonnes of CO_2/toe, 4.1 for coal, 3.1 for oil and 2.4 for gas. The ultimate reserves (U) for each fossil fuel are shown. The right scale in Gtoe corresponds to the equivalence between CO_2 and Gtoe, for cumulative fossil fuel production. The unusual shape of the curve for coal reflects: 1) the sharp decline of the Chinese production since 2013; 2) the uncertainty of the future of world coal production

Figure 72 shows the expected evolution of the cumulative world production of fossil fuels in Gtoe for an ultimate of 1370 Gtoe as estimated table 12, as well as the projected evolution of associated CO_2 emissions. By 2100, the quantities consumed should be about 1100 Gtoe, and the corresponding CO_2 emissions of 3520 Gt (960 GtC). By 2200, CO_2 emissions from fossil fuels should be about 4150 Gt (1130 GtC).

For a long time, the IPCC used the so-called SRES (Special Report on Emissions Scenarios) scenarios to predict the evolution of the Earth's surface temperature in relation to greenhouse gas emissions. In their overwhelming majority, these scenarios predict a continual increase in fossil fuel emissions, which implies an unceasing increase in the annual consumption of fossil fuels. It is remarkable that these scenarios

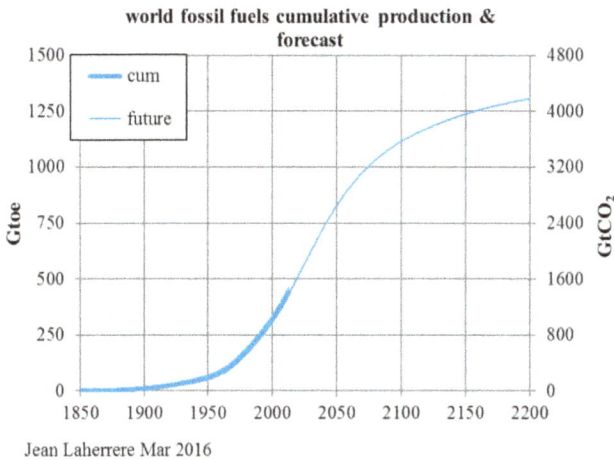

Jean Laherrere Mar 2016

Figure 72 *Cumulative production of fossil fuels from 1850 to present, in Gtoe of contained energy, and associated CO_2 emissions, in Gt, and projected to 2200.*

do not include those of the energy agencies or of the ASPO! Then, these are under the vast majority of the SRES scenarios.

This is not surprising, as the SRES are in fact politico-economic scenarios of future demand, built by a group of economists from the Vienna-based International Institute for Applied Systems Analysis (IIASA). They have been confronted neither with the possible geological supply nor even apparently with the demand foreseen by the agencies.

This critique does not in itself call into question the physical basis of current climate models. It simply says, and this has been repeated to their authors for years, particularly by members of ASPO without any success so far, that the CO_2 emissions scenarios that they use as input data are no more than politico-economic scenarios, "story tellings" having their internal logic, but which does not take into account either the physical feedbacks resulting from the geological limits or even those resulting from the modifications of the climate! In a recent publication, an analysis by *Mohr et al. (2015)* goes in exactly the same direction.

In recent years, however, the IPCC no longer refers to SRES scenarios, but to scenarios known as Representative Concentrations Pathways (RCP), which are *a priori* scenarios of emission trends to describe the changes in the climate and in particular the radiative forcing and global average surface temperatures which would occur if these emissions followed the trajectories described in these scenarios. They are indexed according to the radiative forcing that would be observed in 2100, in W/m^2 of earth's surface: RCP 2.6; RCP 4.5; RCP 6 and RCP 8.5. There is therefore no longer any political-economic justification but only a system of abacuses allowing an assessment of the climatic changes caused by an increase in the greenhouse gas concentrations of the earth's atmosphere by comparing the trajectory emissions levels actually observed and those of the baseline scenarios. They have recently been prolonged from 2100 to 2500 by extended concentration pathways (ECP), which are scenarios assuming *a priori* shapes for the evolution of CO_2 emissions after 2100.

But RCP scenarios are no more physically justified than the SRES scenarios, of which they are in fact a selection.

It is curious that no climate scientist seems to have yet perceived the internal contradiction in the RCP scenarios used by the IPCC to make its climate change expectations. If we do nothing, says the IPCC, we risk following RCP 8, 5, and then it will be a human and economic catastrophe. However, if there is a human and economic catastrophe, we will not be able to follow the scenario RCP 8.5, since it presupposes an ever-growing economy fuelled by fossil fuels produced on demand and therefore not disturbed by the changes of the climate!

That being said, scenario RCP 8.5 is in any case unrealistic (Figure 73). Most of the CO_2 emissions come from fossil fuels. However, their ultimate reserves, according

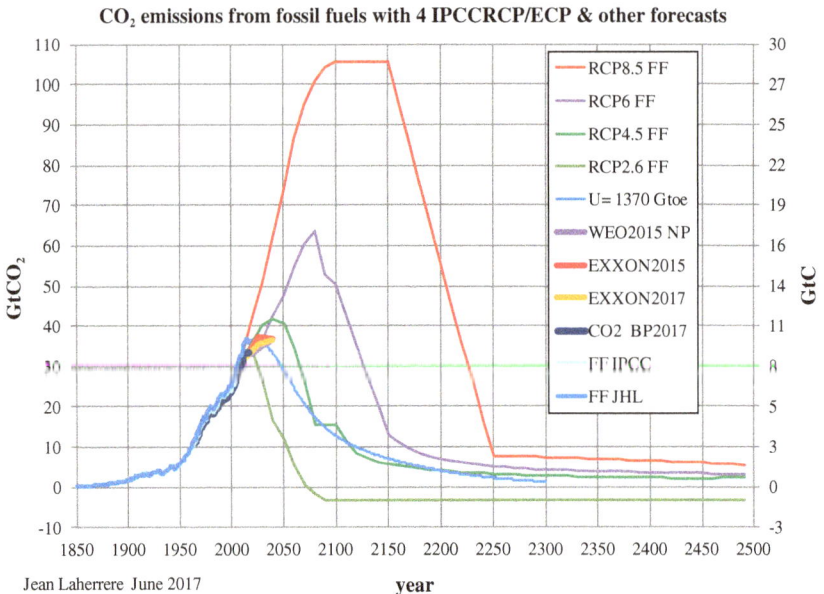

Figure 73 Comparison of cumulative CO_2 emissions from fossil fuels according to Jean Laherrère (FFJHL), in blue, with those of the RCP/ECP scenarios of the ICCP. It can be seen that Jean Laherrère's forecasts are somewhat lower than those of the RCP 4.5, but above RCP 2,6. In this perspective, the RCP 8.5 scenario is totally unrealistic and the RCP6 scenario is highly unlikely. Also note that predictions by the Exxon company are also very far from RCP 8,5 and RCP 6.

to fossil fuel geologists, are 3 to 4 times lower than those implied by this extreme scenario of the IPCC. The RCP 6 scenario is very unlikely, while the RCP 4.5 scenario is closest to the most probable scenario on a geological basis, while on average a little above.

There is undoubtedly a considerable margin of uncertainty in the expectations of J. Laherrère which are used in the above, in particular because of the inevitable inaccuracies in the estimation of ultimates of fossil fuels, in the first place for coal. However the deviations from the IPCC's expectations in a business as usual scenario (BAU), if nothing is done to reduce them, are so important that this poses a considerable problem: Reference is made to pages 57 and 58 of the 2013 IPCC Working Group III (WG III) report, Mitigation of Climate Change, which states that the BAU scenario adopted by this group anticipates CO_2 emissions from fossil fuels between 50 and 75 $GtCO_2$ in 2050 and up to 90 $GtCO_2$ in 2100, while Jean Laherrère's expectations are only 32 and 14 $GtCO_2$ on those dates!

These IPCC scenarios, which take into account physical feedbacks in climate phenomena, therefore do not take into account the feedback between climate and the economy, nor the feedback between the production of fossil fuels and the constraints of the geology of the deposits.

If climate physics is increasingly being taken into account by climate scientists, the IPCC seems to have not really thought about the physical meaning of the future emission scenarios it uses to make its predictions.

On the other hand:

- No probability can therefore be affected by the IPCC to the RCP scenarios. The media naturally take advantage of this situation to speak only of the potentially most disastrous scenario and thus create undue stress in public opinion.
- It is impossible from this point of view to define well-targeted public prevention policies, since it is not known on which scenario it is most effective to rely.

It may also be noted, as already noted in Chapter 3.2.5 in connection with Figure 60, that the IEA's 450 ppm scenario, presented by this agency as a scenario in which humanity is compelled by vigorous measures to limit CO_2 emissions of fossil fuels not to exceed 2 °C of temperature increase is in fact close to the scenario anticipated by Laherrère, which is a BAU scenario without constraints other than geological constraints. This is due to the fact that the IEA's BAU fossil fuel production scenarios, although not as exaggerated as the BAU reference by the IPCC WG III, are still very "optimistic", as we have pointed out to several times.

Figure 74 is taken from the Executive Summary for the 2013 IPCC Report. According to this Figure, there would be a near linear relationship between the increase in temperature of the earth's surface by the end of the century and the total quantities of carbon dioxide of anthropogenic origin that will have been emitted since 1870.

If we want to remain at the end of the century under the 2 °C increase in temperature, as advocated at the Copenhagen climate conference in 2009, we should not issue from 1870 to 2100 more than 3050 Gt of carbon dioxide (830 GtC of carbon content).

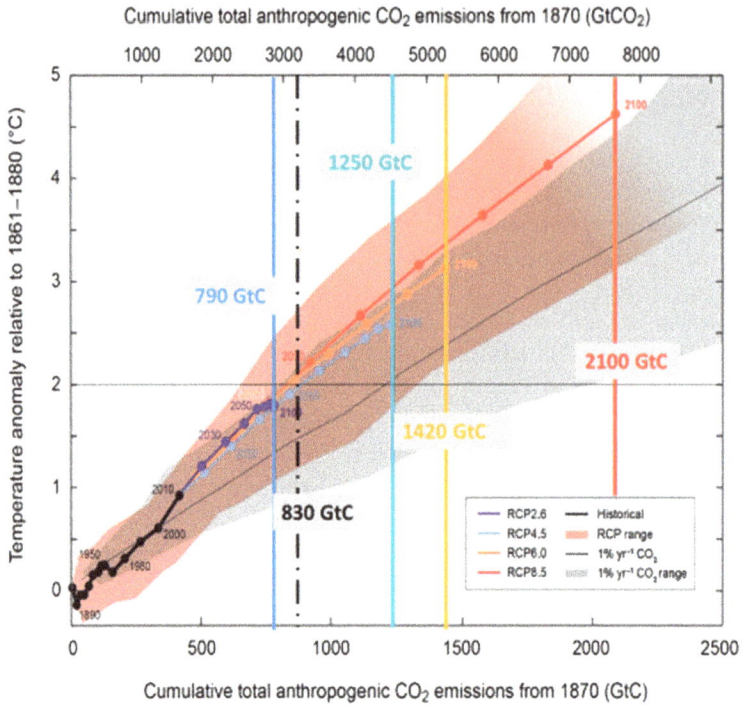

Cumulative total anthropogenic CO_2 emissions from 1870 ($GtCO_2$)

Temperature anomaly relative to 1861–1880 (°C)

1250 GtC

790 GtC

2100 GtC

1420 GtC

830 GtC

RCP2.6 — Historical
RCP4.5 — RCP range
RCP6.0 — 1% yr⁻¹ CO_2
RCP8.5 — 1% yr⁻¹ CO_2 range

Cumulative total anthropogenic CO_2 emissions from 1870 (GtC)

| Figure 74 | *Relation, according to the IPCC 2013 (Source: Climate Change 2013, Science, Summary for Decision Makers, Figure SPM 10), between the increase in surface temperature from 1870 to the end of the century and the amounts of anthropogenic carbon dioxide emitted during that period. The total quantities of carbon dioxide emitted according to the four RCPs are indicated. Also indicated in dotted lines are the quantities not to be exceeded during this period in order to remain below 2 °C of temperature increase.* |

The ultimate production of fossil fuels according to the estimate derived from the scenarios of J. Laherrère corresponds (Figure 72 and Table 15) to about 4670 $GtCO_2$ in round count, i.e. about 1270 GtC in contained carbon, of which 725 GtC for coal, 330 GtC for oil and 215 GtC for gas. However, only about 960 GtC would be emitted from 1870 to 2100, which is considerably less than the corresponding emissions of RCP 4,5 (1,250 GtC), and 1,130 GtC from 1870 to 2,200 (Figure 72). According to Table 12, future emissions corresponding to the remaining reserves are about 860 GtC (Table 15).

The scenario RCP 8.5 is therefore completely unrealistic and should therefore be removed from the IPCC references. The RCP scenario 6 is highly unlikely, but given the uncertainties, especially on the ultimate coal reserves, can be extreme rigor taken as the upper limit of the possible.

Table 15 *Ultimate reserves and reserves yet to be produced of fossil fuels, in Gtoe according to the most probable scenario of ASPO, and corresponding CO_2 emissions, in $GtCO_2$ and GtC. The conversion coefficients are 3.1 tCO_2/toe for petroleum, 2.4 tCO_2/toe for gas, and 4.1 tCO_2/toe for coal.*

Fossil fuel	Oil	Gas	Coal	Total
Ultimate reserves (U), Gtoe	390	330	650	1370
Ultimate CO_2, Gt CO_2 (GtC)	1209 (329)	792 (216)	2665 (727)	4666 (1272)
Reserves yet to be produced, Gtoe	210	237	473	920
CO_2 yet to be produced, $GtCO_2$ (GtC)	651 (178)	569 (155)	1939 (529)	3159 (862)

In order to stay under the fateful 830 GtC by 2100, it would certainly take a major effort, especially since RCP scenarios 2.6 and 4.5 assume total anthropogenic emissions of carbon dioxide, not limited to fossil fuel[39]. But this effort, which is expected to be a reduction of 130 GtC by 2100 for fossil fuels if it is assumed that they would continue to represent 85% of total anthropogenic CO_2 emissions, stay in the limits of the reasonable, which would not be the case with scenario RCP 8,5.

It is immediately apparent that, according to the estimates considered geologically most probable by J. Laherrère, it is the coal which, with about 1940 Gt of future CO_2 emissions (529 GtC), or 63% of the emissions of fossil fuels to come if nothing is done, will be the main risk to the climate. It is therefore on its reduction that priority should be given to the effort. Yet 90% of coal consumption is currently owed to 11 countries, in order of importance China, the United States, India, Japan, the Russian Federation, South Africa, Germany, South Korea, Poland, Indonesia and Australia, which have a special responsibility here.

It would be a good thing for the international community to remind them vigorously, and it would be prudent to organize in these countries a rapid decline in the use of coal in electricity generation, which now accounts for 70% of world coal consumption. A moratorium on the construction of new coal-fired power plants is therefore highly desirable, but failing which, a very large increase in the price to be paid for the emission of carbon dioxide is necessary.

But it will be particularly difficult to convince those on this list whose development depends for the time being primarily on the increase of this consumption, South Africa, China, India, Indonesia. The same is true of other countries which do not appear on this list, but which depend heavily on coal, for example Vietnam. They will not be able to resolve this dilemma until they have a significant supply of substitutes to coal, nuclear and renewable energy, for their electricity generation. It will also be difficult to convince the large exporters, Australia, Indonesia and Colombia, who derive a very significant source of income from coal exports.

---

[39] Most emissions other than fossil fuels are due to deforestation and changes in land use. Can we hope that they will soon become negligible?

A very often mentioned possibility is also to equip coal-fired power plants with carbon capture and storage devices (CCS): carbon dioxide is captured in the fumes of power stations and then stored (sequestered) in geological structures likely to retain it for at least a thousand years. However, there are many obstacles to achieving the CCS, including in particular the investment cost, the reduction in the electrical efficiency of the power plants thus equipped (about 30% more coal would have to be consumed to produce the same amount of electricity), and the reluctance of populations to accept geological storage. It is therefore very doubtful whether the CCS can develop rapidly enough to solve the problem posed!

6

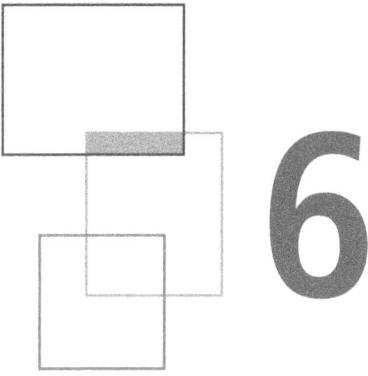

Fossil fuels and public health

The extreme focus of the media on the greenhouse effect and nuclear power leads them to almost completely obscure the direct dangers of using fossil fuels. The citizens of the Western countries have thus remained more or less to the accounts of Emile Zola and Upton Sinclair on the dangers of working in the coal mines in the 19th century and realize very little the present situation.

In fact, fossil fuels have never caused more damage than at the beginning of the 21st century, including in Europe.

These injuries can be classified into two main categories: 1-accidents and 2-diseases, mainly broncopulmonary and cardiovascular, due to pollution outdoor and indoor of the atmosphere by the products of their exploitation and their combustion:

For accidents, the magnitude of a recent estimate at the world scale *(Durand, 2014a)* is 10,000 deaths per year in coal mines and their annexes and 5000 deaths per year due to oil and natural gas. The former mainly concern professionals, coal miners, mainly in the underground mines of Asian countries with large producers, as well as in Russia and Ukraine, where the level of security is currently much lower than in the Western countries. Few people feel concerned in the West. The latter, due to a wide variety of accidents during operation, transport, refining, distribution and use, concern the professionals, but even more the neighbours of the installations and the consumers. Western countries are less affected than others, but are still very much affected, about a quarter of the total mortality due to these accidents in the world. But the lack of media interest in these accidents beyond a few days, apart from a few very spectacular cases, makes them disappear very quickly from the collective memory.

That being said, accidents account for less than 0.5% of fossil fuel mortality, because the bulk of it is due to the products of their combustion. In particular, outdoor and indoor pollution of the atmosphere is caused by very fine particles, but also by oxides of sulfur and nitrogen and a variety of other toxic products, particularly in the case of coals of hazardous elements at low doses, such as arsenic, cadmium, fluorine, mercury, selenium and thallium (see first part, Chapter 1.3) This pollution extends to soils and waters, affecting fauna and flora, including crops. It is also responsible for degradation of buildings.

From the data published in 2014 by the World Health Organization (WHO), the study cited above deduces that fossil fuel combustion products account for between 3 and 4 million premature deaths per year worldwide, by pulmonary and cardiovascular diseases but also by cancers, of which about half are due to coal. The heaviest tribute is paid by the very large coal consumers of the Asian countries, first of all China and India. These figures have been confirmed by the WHO 2016 report.

To this must be added the mortality of miners of coal caused by pulmonary diseases due to the particles suspended in the atmosphere of the mines, the first of which silicosis, about 500,000 premature deaths per year, especially in the countries producing coal by underground mines, China and India, but also there again Russia and Ukraine.

Therefore the overall toll of coal on human health is of the order of 2 millions premature deaths a year.

In Europe 28, the order of magnitude emerging from the European Commission's recent reports is about 300,000 to 350,000 premature deaths a year, mainly due to the use of fuel by the residential and tertiary sector[38], to vehicles exhausts, and to the use of coal, hard coal or lignite, which are used in particular for the production of electricity in a number of European countries, mainly Germany, Poland, the United Kingdom, the Czech Republic, Balkanic countries and Denmark.

However, these figures need to be put into perspective, as they are not as rapid deaths as accidents, but premature deaths, in fact a decrease in life expectancy in affected populations compared to non-affected populations (*Durand*, *2014b*). But this decrease can be of the order of ten years for a very significant part of the affected populations.

---

[38] Let us also point out the important danger of atmospheric emissions from wood heating, especially when used in open fireplaces, as well as without an electrostatic filter, which is practically everywhere.

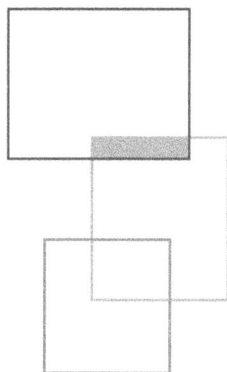

Conclusion

Although they all have the same origin, i.e. a "leak" from the biological cycle of carbon towards the sedimentary cycle, the exploitable accumulations (deposits) of fossil fuels are none the less of great variety and complexity, depending on the nature of the biomass that gave birth to them, the physicochemical conditions of their formation and the geology of their reservoir. However, fossil fuel geologists and geochemists now know how to unravel the threads of their complex histories and model their formation.

This overview gave an idea of the main categories of fossil fuels and of the mechanisms of formation of their deposits. It also provided an overall picture. This allows a better understanding of the possibilities they offer for the production of energy, and of the issues that are attached to them.

Just as a living organism cannot exist without the energy it takes from its food, no human society can survive without the flow of primary energy it extracts from natural sources. The more this society produces material goods, the greater this flow of energy must be.

At present, the flow of primary energy that gives life to human societies is made up of more than 80% of the energy extracted from fossil fuels. Consumers and economists only want to pay attention to their prices, which are of great importance in everyday life. But it is a rather vain exercise to envisage the future, given the impossibility duly observed to predict them even in the short and medium term. This diverts attention from what is so important for this century because of the great importance of fossil fuels, especially oil, in the current course of the economy of industrialized countries: Their future availability.

The analysis that is made here shows that we should be very seriously concerned, starting with oil: the world production of conventional crude oil, the cheapest to produce, and which still provides for the moment three-quarters of our "all liquids" oil supply comes mainly from a small number of very large deposits mostly discovered between 1950 and 1980. It began to decline some ten years ago. On the other hand, at least in volume, the total world production of all-liquids oil has continued to increase: It is thanks to natural gas liquids (condensates and liquids from natural gas plants (NGPLs)), in fact extracted from gas, as well as unconventional oils (extra-heavy oils and bitumens, shale oils) and synfuels (biofuels, GTL, CTL). However, without North American shale oil, it would hardly have increased since 2011. It is also very doubtful, given the gradual decline in energy contained per unit volume of production of all liquids petroleum, and the increasing amount of energy used in this production, that the net quantity of energy thus made available to the world society has in fact increased since that date.

Optimists, mostly economists, but also most of those who speak for the major oil companies, are confident in the technological creativity that characterizes the human species. They believe an inevitable increase in market prices in the long run will allow significant conventional petroleum discoveries, an increase in its recovery rate, and an increase in the production of unconventional petroleum oils (extra-heavy oil and shale oil), natural gas liquids (condensates and NGPL) and synfuels (biofuels, GTL, CTL). This would in total for a very long time be large enough to offset, as is still the case at the moment, the decline of fields currently exploited for conventional oil. For them, the "success story" of North America's oil shale will obviously extend to the entire world. It is the view of the optimists that is currently conveyed by the media, thus anchoring the opinion in the idea that this is nothing to worry about for at least a generation.

Pessimists, mostly oil geologists, observe that the decline in the remaining 2P reserves of conventional crude oil accelerates despite the discoveries, and that unconventional oil is expensive oil that the world economy will not be able to bear indefinitely the prices. Their reserves are very poorly defined and the investments to produce them are heavy. In the case of shale (source-rock) oil, they predict its decline in the United States in a few years. The extension of this "success story" to the world scale has not yet taken place. In this respect, they recall that the success of the United States is also based on a highly developed infrastructure and an exceptionally powerful and reactive oil industry with great know-how compared to other countries with similar favorable geology, and backed-up by an easy-going banking system. It also relies on the peculiarities of the mining code of this country, the only one in the world to give the ownership of the underground to the owner of the soil. It therefore has an interest in exploitation, which is not the case elsewhere.

On the other hand, they stress that it is not so much the importance of the reserves but the possible speed of their putting into production that regulates the performances of the world economy from one year to the next: The flow rate of the carburettor of a car is more important to regulate its speed than the volume of its tank, and for a thirsty man the possible flow of the cock is more important than the volume of the barrel! Almost all predict around 2020 a peak of the possible production of all liquids oil, or at best a plateau around 100 Mb/d from 2020 to 2030 approximately.

The situation is the same for natural gas, with a lag of about ten years: The peak of natural gas would occur around 2030. Some also think that the shale (source-rock) gas will significantly improve this outlook, thanks to technological progress, an improved economic model, and its expansion at the world scale. It seems that the opportunities here are more open than for oil.

Shale (source-rock) oil and gas are actually the last frontier for the global oil and gas industry. As for coal, the common assertions that it will be able to ensure the energy consumption of an ever-richer humanity for at least the next two centuries are based more on "convictions" than on physical modeling. These convictions are at the moment more and more shaken. *Heinberg and Fridley (2010)* believe, for example, that the ultimate reserves, i.e. the actual recoverable quantities, are much lower than announced, and that reality is catching up with us. *Rutledge (2011)* confirms this, by "postmortem" studies of coal basins currently practically exhausted, and whose ultimate reserves had initially been considerably overestimated in relation to the reality of subsequent production. This is particularly the case for coal in the United Kingdom, the initial fuel of the Industrial Revolution. According to *Rutledge (2011)*, around 90% of the global ultimate reserves of coal will have been extracted in 2070. *Friedley (2012)* warned about that China, by far the world's largest producer of coal, was reaching its peak. Indeed the Chinese production is in strong decline since 2013. According to J. Laherrère, this could be already the global peak coal! However on a global scale, uncertainties are more important for coal than for oil and gas.

Pyrolysis of bituminous shale, underground gasification of coal and gas hydrates are commonly cited as sources of fossil fuels, but we are still a long way from being able to move to the massive industrial stage, and also in overcoming their environmental disadvantages. These resources will undoubtedly remain in this century the "resources of the future", as they are today.

These analyzes do not, of course, proceed from an exact science: we have seen the great uncertainties and the wide disagreements that exist on the estimates of the remaining reserves of fossil fuels and those of their future production rates. One can always hope for unexpected discoveries of deposits of unknown type, or technological revolutions. It is no doubt necessary to perfect the analyzes and models with which one seeks to predict possible productions in the future, but also to create economic models that link these productions to the progress of the economy. It is very surprising, given the major importance of the subject for equilibrium of human societies, that so little reflection is currently devoted to it by economists.

We must reason in terms of probabilities: Formally, it is not completely impossible that improving of exploration and production techniques will result in so much progress that peaks of fossil fuel will be delayed for decades. However, the probability of a decline in the possible supply of primary energy supplied by fossil fuels in the next decade looks much higher, and an increase in its demand under the effect of the population growth and also the aspiration of many populations to more well-being. This suggests many turbulences in the global economy and society.

Future carbon dioxide emissions from the use of fossil fuels are essential inputs to climate models. Without questioning the physical basis of these models, however, the quality of the emission estimates that have so far been used by the IPCC can be questioned, as they have not been made on a physical basis, but politico-economic one. This is very damaging for the definition of public policies for climate protection, which use these estimates. Among Representative Concentration Pathways (RCP) scenarios, which are currently the most popular scenario, RCP scenario 8.5 is unrealistic and RCP scenario 6 very unlikely. The geologically most likely scenario should be somewhat below RCP 4.5 and this would represent an increase in temperature from 1870 to 2100 according to the IPCC criteria of between 2 and 3 °C. However, many climatologists recommend not to exceed 2 °C, or even 1.5 °C, but the effort required to meet these criteria is of course much less than with scenarios such as the RCP 6 scenario and even more the RCP 8.5, presented too often by the media as the future if we do nothing!

Ultimately, it is the consumption of coal that is the main threat to the climate, if the criteria of climatologists are taken into account, rather than consumption of oil or gas. In a few years it might again become as during the 19th century and during the 20th century until 1965, the main source of primary energy of mankind. If no other constraint is imposed on it, as is currently the case, about 60% of the future carbon dioxide emissions from fossil fuels might be due to the use of coal.

Large coal consumers (Figure 75) have a great responsibility here because they are the ones who can most effectively constrain climatic changes to remain within acceptable limits by severely restricting their consumption of coal. However, the development of some of them, South Africa, China, India, Indonesia, is currently dependent

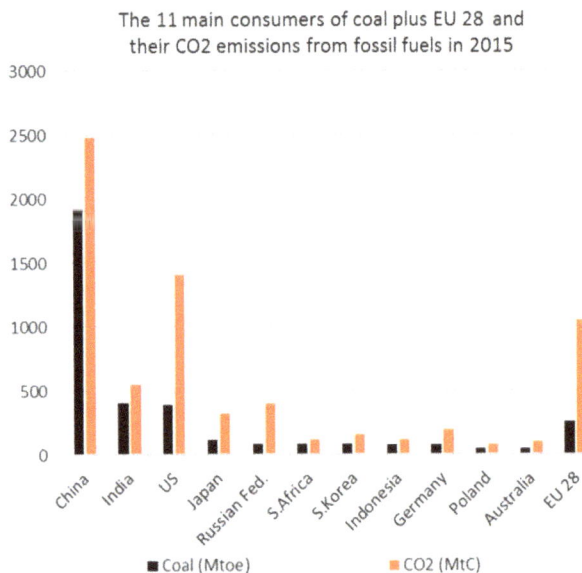

Figure 75   The world's 11 main consumers of coal (in Mtoe) in the world and the EU 28 and their CO_2 emissions (in MtC content) in 2015. Source IEA and BP.

primarily on the increase of this consumption. The same is true of other developing countries that do not appear on this list, such as Vietnam and other Southeast Asian countries. They will not be able to resolve this dilemma until they have sufficiently developed coal substitutes in the production of electricity (nuclear power, renewable energies).

Indeed, the main use of coal, about 70% by mass of its production currently, is the production of electricity. Figure 76 shows for the ten largest electricity producers in the world the large differences in CO_2 emissions associated with this production that exist from one country to another. France is the one of those countries whose electricity production emits the least CO_2 per kWh produced: it owes it to the importance of nuclear power and hydraulics in its electric mix, while most other producers use mainly coal and gas for that. Among other things, it emits about 8 times less CO_2 per kWh than Germany or the United States, 12 times less than China and 18 times less than India!

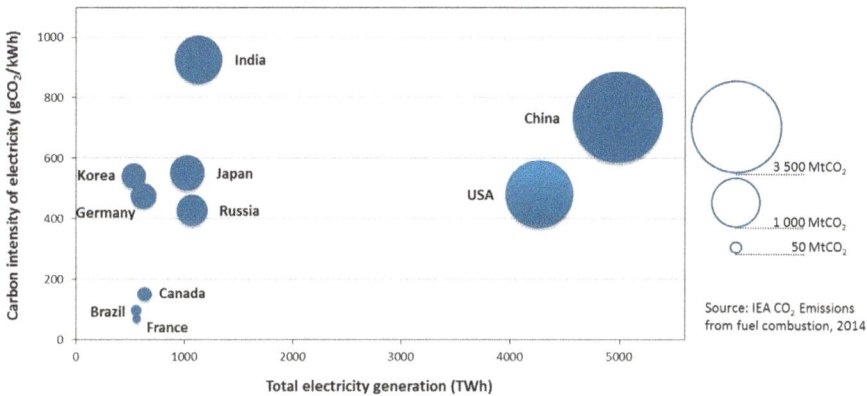

Figure 76 Total CO_2 emissions per kWh of the electricity production of the top ten electricity producers in 2014 and the size of their production. Source: International Energy Agency (IEA), https://www.iea.org/.../2015-04-28-carbon-emissions-from-electricity- generation-for-the-top- ten-producer.html.

The climate problem would thus become much less worrying if these major electricity producers decided to follow the path already long taken by countries like France and Sweden, whose electricity production now makes very little use of coal and more generally fossil fuels, and essentially nuclear and hydraulics. Most of them already have the technical means. The main obstacle is psychological, since it is the fear of a major nuclear accident.

Coal is also that of fossil fuels, and even of all primary sources of energy used by man, which is by far the most important danger to public health, in particular due to air and soils pollution by fine particles, oxides of sulfur and nitrogen, and other harmful elements, such as arsenic, cadmium, fluorine, mercury, selenium, thallium, etc., resulting from its use. This is an order of magnitude of 2 million premature deaths a year, especially in the major consumer countries of Asia, but Europe is not immune! This danger could be considerably reduced if there was a real awareness of this, inducing protection measures far more effective than the current ones. The extreme media focus on the greenhouse effect, renewable energies and nuclear power does not favor it at this time. A serious misunderstanding has thus been created in the public opinion, which only hears about the risks created by greenhouse gas emissions, which do not kill, and those of nuclear, which on the balance sheet killed very little in comparison with coal, which kills enormously. Even alarming reports by Greenpeace *(Greenpeace, 2008)* have received little media attention, which usually make room generously to this organization!

However, in recent years, the risk of pollution by fine particles is beginning to be increasingly evoked *(Durand, 2014b; The Lancet Commission on pollution and health, 2017)*.

With regard to oil, that of fossil fuels which is the more vital for industrial societies, the all liquids quantities produced in the world will likely increase in volume at a slow rate for a few more years, but not in the quantity of energy available to the global society. The volume of oil put on the world market, currently about half of total production, is already declining as a result of the growing domestic consumption of exporting countries (Figure 77). Unless the latter very quickly decide on considerable efforts to save and /or use substitutes (for example nuclear power instead of fuel for their electricity production, as France did after the oil shocks), this trend will be accentuated! First of all, this threatens the industrialized countries without reserves on their soil as are most European countries, but also Japan and South Korea, and soon China and India.

Because of their considerable weight in the functioning of industrial societies and their potential for changing the climate, it is the future evolution of fossil fuel production that will necessarily rule the tempo of the Energy Transition that is being talked about so much.

As we have seen, there is a high probability that the total amount of energy fossil fuels can put each year at the disposal of humanity, all sources being summed up, begins to decline around 2025. Given the increase of the world's population, this decline will be even greater per capita of the planet.

Given the close relationship between the availability of fossil fuels and the progress of the world economy, and the damage caused to it by the oil shocks of 1973 and 1979, an irreversible and rapid decline in the amount of energy we can derive from all fossil fuels is potentially catastrophic. They cannot be managed, given the great inertia of industrial systems, unless very strong countermeasures are taken now. This should therefore very seriously challenge opinion and decision-makers.

Yet, surprisingly, the issue of fossil fuels is not even mentioned in most debates on Energy Transition that are now raging all over the world, which consist mainly of debates on electricity alone, renewables against nuclear.

World imports and exports of crude oil & refined petroleum products

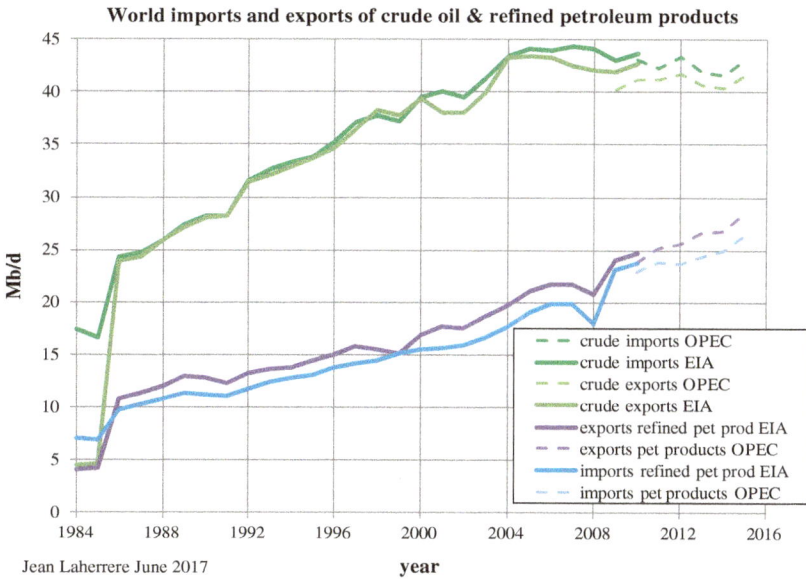

Jean Laherrere June 2017

Figure 77 *Evolution of exports and imports of crude oil since 1984, according to the EIA until 2010 and according to OPEC from 2010 to 2014: they have been declining since 2008. This Figure also shows the evolution of imports–exports of refined products. Note that unconventional petroleum represents little in world trade, and that these crude oil curves are therefore mainly concerned with conventional crude oil, which is therefore the main source of imports–exports.*

However, the best energy policy that can be followed at the world scale, and particularly by industrial countries that have no or very little fossil fuels reserves, such as most European countries, but also Japan and South Korea, is undoubtedly to anticipate the decline of fossil fuels, first and foremost that of oil. The priority of this policy should therefore be to seek to reduce oil consumption as quickly as possible and to find substitutes for it in its main uses, which are mobility and heating. The technical instruments which can be used to follow the mainlines of this policy already exist:

- For mobility: vehicles with very low fuel consumption and/or electric vehicles, electrification of public transport.

- For heating: better thermal insulation of buildings, solar heaters, wood heating, recovery of heat from the ground and the air by means of heat pumps... substituting oil, but also gas, the life expectancy of which does not seem to be much greater than that of oil if the "success story" of shale gas in the United States is not transmitted to the rest of the world[39]. However, it remains to develop these techniques to the necessary level.

---

[39] To do this, it would be necessary to at least modify the mining codes so that the owners of the soil are associated with the profits of the farms, and this is very unlikely in the short-term.

This policy would also be much more effective not only in combating greenhouse gas emissions but also in order to improve the trade balance of countries without significant reserves, which already suffers greatly from the burden of fossil fuel imports, than the current policy of forced march in the development of wind and photovoltaic solar energy[40] which seems to be for the moment the main focus of the "Energy Transition".

---

[40] Wind and solar photovoltaic produce a no-carbon electricity, but this is hardly in itself interesting for countries whose electricity already emits very little CO_2 per kWh produced, such as in Europe, France (Figure 77), but also Norway, Sweden and Switzerland. If the objective is really to fight against CO_2 emissions, it is not on electricity production that the effort must be made in these countries, but on the highly emitting sectors such as transport and heating. Moreover, the materials needed for wind and solar PV are in these countries almost entirely imported, which degrades their trade balance instead of improving it!

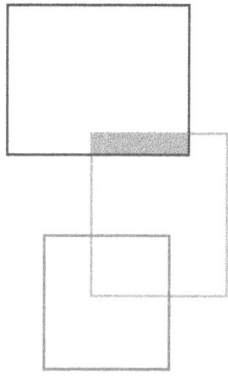

Appendix 1: Chemical composition of oils

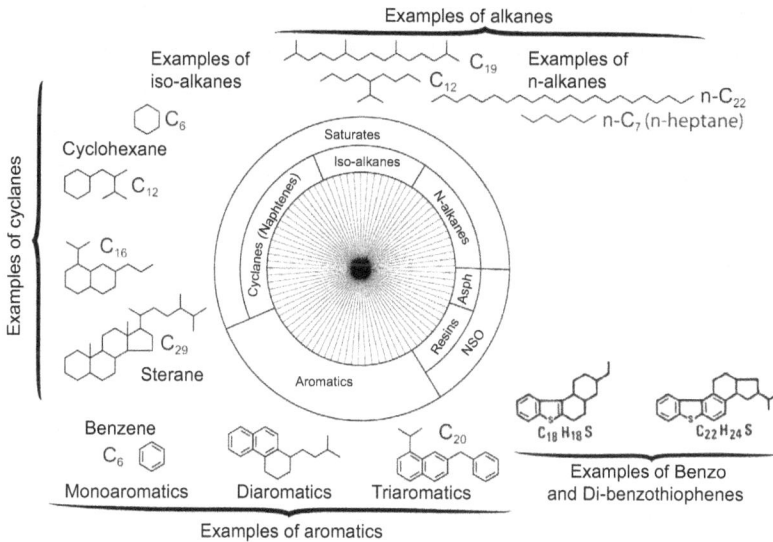

Examples of alkanes

Examples of iso-alkanes

C_6 Cyclohexane

C_{12}

C_{16}

C_{29} Sterane

Examples of cyclanes

Saturates

Iso-alkanes

Cyclanes (Naphtenes)

N-alkanes

Asph

Resins

NSO

Aromatics

C_{12}

C_{19}

Examples of n-alkanes

n-C_{22}

n-C_7 (n-heptane)

Benzene

C_6

Monoaromatics

Diaromatics

C_{20}

Triaromatics

$C_{18}H_{18}S$

$C_{22}H_{24}S$

Examples of Benzo and Di-benzothiophenes

Examples of aromatics

Figure A1 *Proportions of the different families of the main chemical compounds found in a "medium" crude oil and examples of hydrocarbon molecules belonging to these families: Petroleum hydrocarbons belong to two major structural families: saturated hydrocarbons and aromatic hydrocarbons.*

Commentary of Figure A1. A saturated hydrocarbon, also referred to as an aliphatic hydrocarbon, is a hydrocarbon in which each carbon atom is bonded via a single bond to 4 neighboring atoms of carbon or hydrogen. Here only the carbon atoms are shown, the number of hydrogen atoms bonded to each carbon atom being calculated by deducting from 4 the number of carbon atoms bonded to this atom. The total number of carbon atoms for each molecule represented has been reported.

Saturated hydrocarbons include two subfamilies:

- Linear saturated hydrocarbons, also called paraffinic hydrocarbons, paraffins or alkanes. They include straight-chain n-alkanes, and branched chain isoalkanes.

- Cyclic saturated hydrocarbons, called also cycloalkanes, cyclanes or naphthenes. Those comprise at least one saturated ring with 5 or 6 carbon atoms in their structure. The simplest are cyclopentane C_5H_{10} and cyclohexane C_6H_{12}, the latter being shown here.

Aromatic hydrocarbons, also called arenes, are those which comprise from one to three aromatic rings, exceptionally four or even five, also called benzene rings. The benzene C_6H_6, shown here, is the simplest of aromatic hydrocarbons. A benzene ring is characterized by the existence of so-called benzene bonds between its carbon atoms, with electron delocalization. These bonds are represented by a double bond at each two carbon atoms of the ring as here, or by a circle at the center of the hexagon. These aromatic rings are often associated with saturated rings, in so-called naphtheno-aromatic hydrocarbons.

In the hydrocarbons whose structure is represented in these Figures, two biomarkers can be observed:

- the C_{19} isoalkane possessing a CH_3 branch every 4 carbon atoms. It is an iso-prenoid called pristane, derived from the side chain of chlorophyll;

- the C_{29} sterane, which derives from a biological sterol (of which the well-known cholesterol is an example).

This construction set with only 2 elements, carbon and hydrogen, therefore has many subtleties.

Are also shown, examples of benzothiophenes and dibenzothiophenes, sulfur molecules, which in the standard protocol for separation of the main hydrocarbon families (SARA analysis, see legend in Figure A2) are also found in aromatics, whereas, *sensu stricto*, they are not.

Commentary of Figure A2. On the left: supposed structures of resins and asphaltenes: these are large molecules comprising hetero-elements, essentially oxygen, sulfur and nitrogen. Asphaltenes are maintained in pseudosolution in petroleum thanks to the resins which are the "binder" between them and the hydrocarbons (a). It is easy to precipitate asphaltenes in the laboratory by adding light hydrocarbons to a small amount of crude oil. The asphaltene molecules then agglomerate and then precipitate as black particles (b). This is called deasphalting. This phenomenon can occur in nature when light gas or oil invades an oil field where the oil is rich in asphaltenes.

On the right: remarkable biomarkers: a porphyrin (2) and two isoprenoids, phytane and pristane (3), both derived from chlorophyll a (1) side chain. It should be

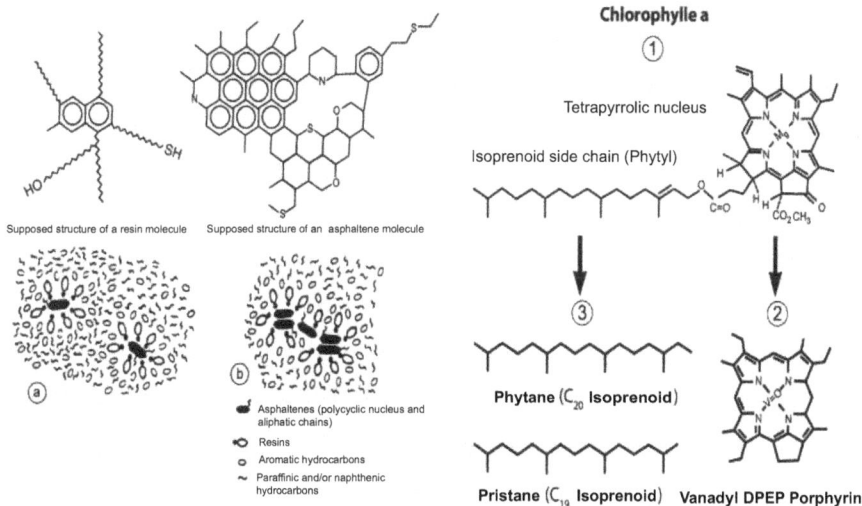

Figure A2   Resins, asphaltenes and biomarker.

noted that in the tetrapyrrole nucleus of the porphyrin, the magnesium ion of the chlorophyll was replaced by a vanadyl ion. The heavy fractions of oils (fuels) thus contain traces of vanadium, but it can also be nickel, which come from the porphyrins they contain. Porphyrins are found in resins but can be isolated.

The geochemical analysis of an oil begins with a distillation to recover the lightest fraction, which can then be analyzed by gas chromatography. The residue is then broken down in its the four major classes of constituents: saturated, aromatics, resins and asphaltenes (SARA analysis). The asphaltenes are separated by precipitation with a light n-alkane, and after filtration and evaporation of the n-alkane, the other three classes are separated by liquid chromatography. Saturated and aromatics are analyzed by gas chromatography and gas chromatography-mass spectrography (GC-MS) coupling. Benzo and dibenzothiophenes are found in the aromatic fraction, but are not hydrocarbons *sensu stricto* since they contain sulfur. Resins and asphaltenes are analyzed for their overall properties. The porphyrins, which are recovered in the resin fraction, are however separable from the resins by liquid chromatography.

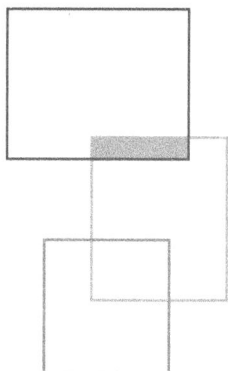

Appendix 2: Dynamics and Resilience of the Production of Shale Oil and Gas

In relation to the production of an oil or gas well from a conventional deposit, a production of shale (source-rock) oil or gas is characterized by:

- A much faster implementation: the decision to drill a well is fast and it takes a few months between this decision and the putting into production. In the case of a newly discovered conventional deposit, drilling can be rapid, but it takes 5 to 10 years to develop all the equipment necessary for production.

- A much more rapid decline in production, from 80 to 90% of initial production in 3 years, against near stability for a few years, followed by a very slow decline for a conventional well (Figures A3 and A4).

- Total production until abandonment much lower, 10 to 100 times less than a well for conventional production.

- A higher marginal cost of production.

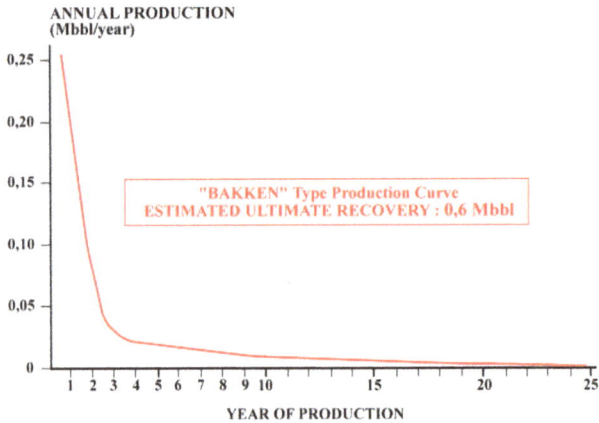

Figure A3 *Average production profile over the years of a shale oil well in the Bakken formation in the United States: after three years, production is only 10% of the initial production. Bbl is another name for barrel. Total production after 25 years is only 0.6 Mb. According to Vially et al. (2013).*

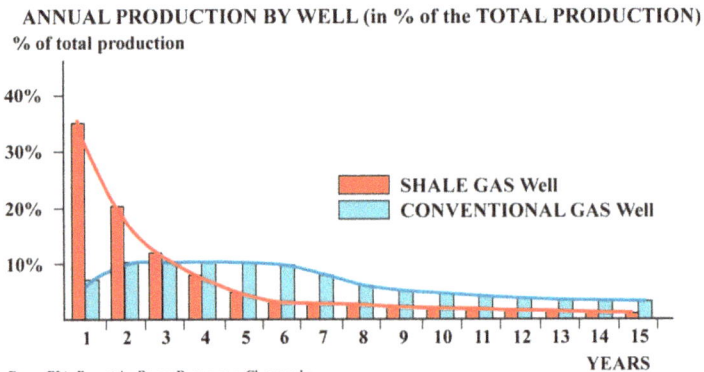

Figure A4 *Average production profile over the years of a shale gas well in the United States and comparison with that of a conventional gas well as a % of total production. Source: Boston Consulting Group.*

The result is a much faster development of oil and shale gas than conventional oil and gas, but also a much greater sensitivity to market prices. It can be seen that if the period of very high oil prices has allowed a shale oil boom in the United States, the current low prices are now breaking production and freezing development attempts outside the United States.

This also results in the need to continuously drill new wells to compensate for the rapid loss of production of wells already drilled: as when walking on a conveyor belt in the direction opposite to the motion, stop walking results in going backward.

However, technological advances are making it possible to mitigate this phenomenon by making better use of existing wells. But the occupation of the soil by the drilling and production facilities is still very high and growing.

This high production dynamics also results in a strong resilience: production can resume very quickly once prices have become profitable again. On the other hand, once these very many wells are drilled, production of the whole can remain at a noticeable level for a long time.

Nevertheless, these productions are limited by the geology of the formations and the decreasing yield of the production of the most favorable zones (sweetspots) over time in spite of the technological progress: They will necessarily pass through a peak, probably very close in the United States, while as yet no very significant production has taken place outside the United States.

It is, after all, only oil and gas contained in reservoirs of extremely low permeability, whose production must be stimulated by energetic means.

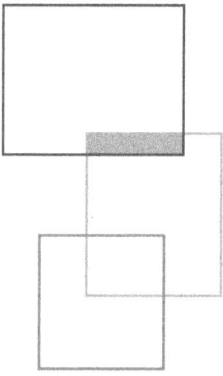

References

Aucott, M. L., Melillo, J. M., 2013. A preliminary energy return on investment analysis of natural gas from the Marcellus shale. *Journal of Industrial Ecology*, 17 (5), 668-679.

Ayres, R., 1998. *Turning Point: An End of the Growth Paradigm*. Earth Scan, 1998 - Business and Economics, London, 258 p.

Belin, S., 1992. Distribution microscopique de la matière organique disséminée dans les roches-mères. Technique d'étude. Interprétation des conditions de dépôt et de diagenèse, thèse de doctorat de l'Université d'Orsay, 371 p.

Bonijoly, M., *et al.*, 1982: A possible mechanism for natural graphite formation. *International Journal of Coal Geology*, 1(4), 263-362.

Bordenave, M. L., 1993. The sedimentation of organic matter. *In: Applied Petroleum Geochemistry*, M.L. Bordenave Ed., Editions Technip, Paris, 15-73.

Bundesanstalt für Geowissenshaften und Rohstoffe (BGR), 2013. Energy Study 2013, reserves, resources and availability of energy resources (17), Hannover, 112 p.

Burrus, J., 1997. Contribution à l'étude du fonctionnement des systèmes pétroliers, apport d'une modélisation bidimensionnelle, thèse de doctorat en sciences, spécialité hydrologie et hydrogéologie quantitative, tome 2, Ecole des Mines de Paris, 346 p.

Campbell, C., Laherrère, J., 1998. The End of Cheap Oil: global production of conventional oil will begin to decline sooner than most people think, probably within 10 years. *Scientific American*, March 1998, 78-83.

Capen, E.C., 1996. A Consistent Probabilistic Definition of Reserves. *SPE Res. Eng.* 11(1), 23–28. SPE-25830-PA. http://dx.doi.org/10.2118/25830-PA.

Caruana, J., 2016. Credit, commodities and currencies. Lecture at the London School of Economics and Political Science, London, 5 February 2016. https://www.bis.org/speeches/sp160205.pdf.

Charlez, P., 2016. Geopolitics of Unconventional Resources outside North America. SPE Annual Technical Conference and Exhibition held in Dubai, UAE, 26–28 September 2016.

Chavanne, X., 2015a. Définition des hydrocarbures de pétrole et gaz. http://aspofrance.viabloga.com/files/DefHClong-XCmrs15.pdf.

Chavanne, X., 2015b. L'efficacité de la filière pétrolière. http://aspofrance.viabloga.com/files/EffEnergieFilierePetrole-XCjuin15.pdf.

Chavanne, X., 2016. Production des pétroles naturels 2000-2025. https://aspofrance.files.wordpress.com/2016/10/xc-petrolesnaturels-juin16.pdf.

Clayton, C., 1992. Source volumetrics of biogenic gas generation. *In: Bacterial gas*, R. Vially ed., Editions Technip, 191-204.

Connan, J., 1984. Biodegradation of crude oils in reservoirs. *In: Advances in petroleum geochemistry*, J. Brooks & D.H. Welte Eds., Academic Press, London, 299-335.

Craig, J., 2015. Global Climate, the Dawn of Life and the Earth's Oldest Petroleum Systems. *Société géologique de France*, Paris, Novembre 2015http://geosoc.fr/adherent/s-inscrire/cat_view/109-conferences/115-2015/116-les-roches-meres-petrolieres.html

Demaison, G. J., Moore, G. T., 1980. Anoxic environments and oil source bed genesis, *Bull. Am. Assoc. Petr. Geol.*, 64, 1179-1209.

Deffeyes, K., 2001. *Hubbert's Peak: The Impending World Oil Shortage*, Princeton University Press.

Durand, B., *et al.*, 1980. *Kerogen, insoluble organic matter from sedimentary rocks*, Editions Technip, Paris, 519 p.

Durand, B., 1988. Understanding of HC migration in sedimentary basins (present stage of knowledge). *Org. Geochem.*, 13, 445-459.

Durand, B., 2003. A History of Organic Geochemistry. *Oil & Gas Science and Technology-Rev. IFP*, 58(2), 204-231.

Durand, B., 2014a. Les dangers des combustibles fossiles (autres que l'effet de serre). http://www.sauvonsleclimat.org/universite-dete-2014-bordeaux.

Durand, B., 2014b. Les risques de la pollution atmosphérique pour la mortalité prématurée. Comparaison avec ceux du tabac et ceux de la radioactivité. http://www.sauvonsleclimat.org/les-risques-de-la-pollution-atmospherique-pour-la-mortalite-prematuree-comparaison-avec-ceux-du-tabac-et-ceux-de-la-radioactivite/35-fparticles/1681-les-risques-de-la-pollution-atmospherique-pour-la-mortalite-prematuree-comparaison-avec-ceux-du-tabac-et-ceux-de-la-radioactivite.html.

Fridley, D., *et al.*, 2012. Review of China's Low-Carbon City Initiative and Developments in the Coal Industry. Ernest Orlando Lawrence Berkeley National Laboratory. https://www.osti.gov/scitech/servlets/purl/1171744

Gagnon, N., Charles, A., Hall, S., Brinker, L., 2009. A Preliminary Investigation of Energy Return on Energy Investment for Global Oil and Gas Production, *Energies*, 2(3), 490-503; doi:10.3390/en20300490.

Georgescu-Roegen, N., 1971. The Entropy Law and the Economic Problem. Traduction française : la loi de l'entropie et le problème économique. *In : La Décroissance*, Nicholas Georgescu-Roentgen. Editions Sang de la Terre 2011. 302 p.

Giraud, G., Kahraman, Z., 2014. How Dependent is Growth from Primary Energy? The Dependency ratio of Energy in 33 Countries (1970-2011). Documents de travail du Centre d'Économie de la Sorbonne 2014.97 - ISSN : 1955-611X. 2014. <halshs-01151590>.

Goldstein, T. P., Aizenshtat, Z., 1994. Thermochemical sulfate reduction a review. *Journal of Thermal Analysis* (1994) 42, 241. doi:10.1007/BF02547004. Greenpeace, December 2008: The true cost of coal. http://www.greenpeace.org/international/Global/international/planet-2/report/2008/11/cost-of-coal.pdf .

Hall, C., Lambert, J., Balogh, S., 2014. EROI of different fuels and the implications for society. *Energy Policy*, 64, 141-152.

Heinberg, R., Fridley, D., 2010. *The End of Cheap Coal*. Nature 468, 367-369.

Höök, *et al.*, 2013. Decline and depletion rates of oil production: a comprehensive investigation. *Philosophical transaction of the Royal Society A,* 2 December 2013. DOI: 10.1098/rsta.2012.0448.

Hovland, M., Irwin, H., 1992. Habitat of methanogenic carbonate cemented sediments in the North Sea. *In: Bacterial Gas*, R. Vially Ed., Editions Technip, 157-172.

Huc, A.-Y., 1980. Origin and formation of organic matter in recent sediments and its relation to kerogen. *In : Kerogen, insoluble organic matter from sedimentary rocks*, B. Durand Ed., Editions Technip, Paris, 339-383.

Huc, A.-Y., Van Buchem, F. S. P., Colletta, B., 2005. Stratigraphic Control on Source-Rock Distribution: first and second Order Scale. *In: Deposition of Organic-Carbon-Rich Sediments : Models*, SEPM Special Publication n° 82, N.B. Harris Ed., 225-242.

Hughes, J. D., 2014. Drilling deeper, a reality check on U.S. Government forecasts for a lasting tight oil and shale gas boom. Post Carbon Institute. http://www.postcarbon.org/wp-content/uploads/2014/10/Drilling-Deeper_FULL.pdf.

Intergovernmental Panel on Climate Change (IPCC), AR 5, 2013: Working Group III-Mitigation of Climate Change, December 2013, Chapter 7, 57-58.

Kümmel, R., 2011. *The Second Law of Economics: Energy, Entropy, and the Origins of Wealth*. Springer, Berlin.

Laherrère, J.H., 2009. Update on US GOM: Methane Hydrates, TOD July 17. http://europetheoildrum.com/node/5552#more.

Laherrère, J., 2011a. Backdating is the key. http://aspofrance.viabloga.com/files/JL_ASPO2011.pdf.

Makogon, Y.F., 2010. Natural gas Hydrates-A promising Source of Energy. *Journal of Natural Gas Science and Engineering*, 2, 49-59.

Martin-Amouroux, J.-M., 2008. *Charbon, les métamorphoses d'une industrie, la nouvelle géopolitique du XXIème siècle.* Editions Technip, Paris, 420 p.

Masset, J.-M., 2009. Pétrole, gaz : Pic ou plateau ? *BRGM, 10 enjeux des géosciences, Dossier spécial année internationale de la planète terre,* 16-26.

Mathieu, Y., 2011. *Le dernier siècle du pétrole ? La vérité sur les réserves mondiales.* Editions Technip, Paris, 138 p.

Mohr, S. H., *et al.*, 2015. Projection of world fossil fuels by country. *Fuel,* 141(1), 120-135.

Muschalik, M., 2017. Brent Exit http://crudeoilpeak.info/brent-exit.

Nelder, C., 2013. Are Methane Hydrates really going to change geopolitics? http://www.theatlantic.com/technology/archive/2013/05/are-methane-hydrates-really-going-to-change-geopolitics/275275/

Randolph, Ph., 1977. Natural Gas From Geopressured Aquifers? SPE Annual Fall Technical Conference and Exhibition, 9-12 October, Denver, Colorado. Society of Petroleum Engineers. SPE- 6826-MS http://dx.doi.org/10.2118/6826-MS.

Rice, D.D, 1992. Controls, Habitat, and Resource Potential of Ancient Bacterial gas. *In: Bacterial Gas,* R. Vially ed, Editions Technip, 91-118.

Rojey, A., *et al.*, 1997. *Natural Gas- Production, Processing, Transport;* Institut Français du Pétrole Publications, Editions Technip, 1997.

Rouxhet, P.G. *et al.*, 1980. Characterization of kerogens and of their evolution by infrared spectroscopy.*In : Kerogen, insoluble organic matter from sedimentary rocks,* B. Durand Ed., Editions Technip, Paris, 164-190.

Rutledge, D., 2011. Estimating long-term world coal production with logit and probit transforms, *International Journal of Coal Geology*, 85, 23-33. http://www.its.caltech.edu/ rutledge/DavidRutledgeCoalGeology.pdf

Sokolov, V., 1974. *Géochimie des gaz naturels,* traduction française, Editions Mir, Moscou, 365 pages.

Stach, E., *et al.*, 1982. *Stach's textbook of coal petrology*, 3rd edn. Gebrüder Borntraeger. 535.

Stiglitz, J., 1974. Growth with Exhaustible Natural Resources: The Competitive Economy, *Review of Economic Studies*, 41(5), 139-152.

The *Lancet* Commission on pollution and Health, 19 Oct.2017 http://www.thelancet.com/journals/lancet/article/PIIS0140-6736(17)32345-0/fulltext

Tissot, B., 1969. Premières données sur les mécanismes et la cinétique de la formation du pétrole dans les bassins sédimentaires. Simulation d'un schéma réactionnel sur ordinateur. *Oil&Gas Science and Technology-Rev.IFP*, 24, 470-501.

Tissot, B. P., Welte, D. H., 1978, 1984. *Petroleum formation and occurrence*, first edition, 1978, second and enlarged edition, 1984. Springer Verlag, 699 p.

Ungerer, Ph., Espitalié, J., Behar, F., Eggen, S., 1988. Modélisation mathématique des interactions entre craquage thermique et migration lors de la formation du pétrole et du gaz, *C.R. Acad. Sci. Paris*, Série II, 307(8), 927-934.

Ungerer, Ph., 1993. Modelling of petroleum generation and migration. *In: Applied Petroleum Geochemistry*, M. Bordenave Ed., Editions Technip, 395-442.

Vandenbroucke, M., Bordenave, M.L., B. Durand, 1993. Transformation of organic matter with increasing burial of sediments and the formation of petroleum in source-rocks. *In: Applied Petroleum Geochemistry*, M. Bordenave Ed., Editions Technip, 101-121.

Van Krevelen, D. W., 1993. *Coal (typology, physics, chemistry, constitution)*, third edn. Elsevier.

Verhulst, P. F., 1845. Recherches mathématiques sur la loi d'accroissement de la population, *Nouveaux Mémoires de l'Académie Royale des Sciences et Belles-Lettres de Bruxelles*, 18, 1845, 1-42.

Vially, R., *et al.,* 1992. *Bacterial Gas*, Editions Technip, 242 p.

Vially, R., 2016. Hydrates de méthane- Ressources et enjeux environnementaux. *Techniques de l'ingénieur*, référence BE 8 561, 10 Janvier 2016.

Weissbach, D., *et al*, 2013. Energy intensities, EROIs, and energy payback times of electricity generating power plants. *Energy*, 52(1), 210-221.

World Health Organization (WHO), 2016. http://www.who.int/mediacentre/news/releases/2016/air-pollution-estimates/en/

Wood, Mackenzie, 2017. Vaca Muerta development study 2017, April 2017.

www.ingramcontent.com/pod-product-compliance
Lightning Source LLC
Chambersburg PA
CBHW042310210326

41598CB00041B/7342